# Green Energy and Technology

Climate change, environmental impact and the limited natural resources urge scientific research and novel technical solutions. The monograph series Green Energy and Technology serves as a publishing platform for scientific and technological approaches to “green”—i.e. environmentally friendly and sustainable—technologies. While a focus lies on energy and power supply, it also covers "green" solutions in industrial engineering and engineering design. Green Energy and Technology addresses researchers, advanced students, technical consultants as well as decision makers in industries and politics. Hence, the level of presentation spans from instructional to highly technical.

**Indexed in Scopus**.

**Indexed in Ei Compendex**.

Vahid Vahidinasab • Behnam Mohammadi-Ivatloo
Editors

# Demand-Side Peer-to-Peer Energy Trading

Springer

*Editors*
Vahid Vahidinasab
Department of Engineering
School of Science and Technology
Nottingham Trent University
Nottingham, UK

Behnam Mohammadi-Ivatloo
School of Energy Systems
LUT University
Lappeenranta, Finland

ISSN 1865-3529 ISSN 1865-3537 (electronic)
Green Energy and Technology
ISBN 978-3-031-35232-4 ISBN 978-3-031-35233-1 (eBook)
https://doi.org/10.1007/978-3-031-35233-1



This Springer imprint is published by the registered company Springer Nature Switzerland AG
The registered company address is: Gewerbestrasse 11, 6330 Cham, Switzerland

# Preface

Advancements in Information and Communication Technologies (ICT) have given rise to new concepts in the energy market, such as transactive energy (TE) and peer-to-peer (P2P) markets. Through these markets, consumers and prosumers can trade energy and services in a distributed, secure, and transparent platform with other players in the network, including small-scale users, community/microgrid managers, system operators, and large-scale producers. This allows players at all levels to participate in providing energy and ancillary services while still achieving their own goals, whether it be individual cost reduction, utilizing green energy, or philanthropic purposes such as aiding those experiencing energy poverty. The objective of this book was to collaborate with experts from diverse disciplines that are related to peer-to-peer and transactive energy systems. The authors strongly believe that each discipline holds valuable knowledge that is not commonly shared with other disciplines. By bringing together these experts, the book aims to provide a comprehensive and detailed reference that can be used to tackle the challenges we face in the current energy landscape.

Chapter 1 offers a comprehensive review of academic literature, research projects, and industry practices that explain the concept of P2P energy trading. It covers market design, the role of end-user behavior, challenges and opportunities, social science perspectives, and policies. Different P2P energy trading management methods such as community-based, decentralized, or distributed methods are analyzed in this chapter.

Chapter 2 aims to present the concepts, challenges, and perspectives of P2P and transactional energy trading. This chapter discusses the changes happening in the electricity industry due to climate change, distributed energy resources, and advancements in information and communication technologies. Peer-to-peer trading is introduced as a solution to managing the unpredictability of renewable energy sources. Different local energy market models such as P2P, Social Self-Consumption, and Transactive Energy are discussed, with blockchains emerging as a promising solution for activating smart contracts.

Chapter 3 provides an overview of these approaches and discusses existing research challenges. Various innovative techniques have been introduced to prepare prosumers for energy markets based on policies and regulations, including structural design, optimal participation, and management using technology and artificial intelligence.

Chapter 4 introduces peer-to-peer (P2P) energy trading as a new transactive energy platform where all parties involved, including producers, consumers, and prosumers, interact with each other through a P2P network. This platform enables flat communication between a large number of individual nodes, which is different from conventional market mechanisms. The chapter discusses the advantages and disadvantages of the P2P energy trading approach and explores the use of blockchain technology in P2P energy markets. It also highlights the security and privacy issues of smart energy systems and summarizes the weaknesses and vulnerabilities of the blockchain-enabled P2P energy trading platform.

Chapter 5 discusses the emergence of peer-to-peer (P2P) local electricity markets at the distribution network level, enabled by smart grid technology. The market comprises retailers, who are profit-driven companies that generate or purchase electricity from various sources, and prosumers, who can also generate electricity and participate in the local market. The chapter proposes a competitive market model that allows for bilateral energy trading between all players, and also enables retailers to participate in the upstream market as smart agents. A decentralized algorithm called the primal-dual sub-gradient method is used to clear the proposed market.

Chapter 6 proposes a fully decentralized blockchain-based peer-to-peer token energy trading market for small-scale prosumers. The market allows prosumers to trade energy with their peers using smart contracts and negotiations. The demurrage mechanism is implemented to prevent energy token accumulation and increase market attractiveness. The primal-dual sub-gradient method is used to clear the market in the presence of the demurrage mechanism.

Chapter 7 considers a central shared battery energy storage system (CSBESS) and uses the alternating direction method of multipliers (ADMM) to model the optimization of the P2P energy trading market in a decentralized manner. Peers can maximize their social welfare by using CSBESS to charge and discharge during off-peak and peak consumption times.

Chapter 8 presents a fully decentralized peer-to-peer energy market for small-scale prosumers with energy storage systems. The prosumers can negotiate with local sellers and compensate for energy shortage from the upstream retail market during hours without local generation. Robust optimization is used to deal with energy price uncertainty. The fast alternating direction method of multipliers (ADMM) is used to clear the proposed decentralized peer-to-peer energy trading between local prosumers.

Chapter 9 discusses the evolution of electricity markets and the role of consumers/prosumers in market management. It introduces the concept of multi-energy peer-to-peer (P2P) systems and their basic structures and components, including energy hubs and transactive energy systems. The chapter also reviews some research papers related to multi-energy P2P systems and presents a model and case study to analyze the application of these systems. The exchange of various energy carriers on the demand side is highlighted as an important aspect of modern integrated energy systems.

Chapter 10 discusses the use of peer-to-peer (P2P) energy trading as a solution to high energy costs during peak demand hours in residential areas. The system allows customers to share their electricity-generating capabilities and interconnects peers in a planned residential low voltage network to decrease energy costs. The study examines three buildings, two of which have PV generation and one with a hydrogen energy storage system, using the conditional value-at-risk (CVaR) technique to analyze the system's behavior under uncertain situations.

To conclude, we would like to sincerely thank all of the authors who contributed to this book and also all the reviewers for their insightful observations and helpful comments.

Nottingham, UK — Vahid Vahidinasab
Lappeenranta, Finland — Behnam Mohammadi-Ivatloo

# Contents

# Chapter 1
# Overview of the Peer-to-Peer Transactions and Transactive Energy Concept, Challenges, and Outlook

**Sobhan Dorahaki, Masoud Rashidinejad, and Mojgan MollahassaniPour**

## 1.1 Introduction

In the last few decades, the structure of the energy systems has fundamentally changed by increasing the role of end users in the smart grid energy systems [1]. In fact, end users represent an active role in the demand side of the smart grid energy systems by generating, converting, and storing energy incorporating Distributed Energy Resources (DERs) and storage devices [2]. The world Photovoltaic (PV) market, as one of the most prevalent DERs, is expected to grow by 11% over the next 6 years, whereas the capacity of residential storage systems is predicted to increase from 95 MW in 2016 to 3700 MW by 2025 [3]. Such promising tendency into green energy resources can improve environmental indices as a result of reducing emitted pollution of conventional power plants which can facilitate the path to reaching the Paris Climate Agreement's goals. However, this positive step toward more green and sustainable energy systems will alter the outlook of the energy framework from the technical, social, economic, and environmental standpoints. Hence, implementation of DERs bring diverse challenges to energy systems [4]. Voltage variations, power losses, line congestions, protection scheme changes, and phase unbalance issues can be named as the main technical challenges, and complexity in operation scheduling can be considered as the economic challenge in energy systems with high penetration of DERs. To overcome such challenges,

S. Dorahaki · M. Rashidinejad (✉)
Department of Electrical Engineering, Shahid Bahonar University of Kerman, Kerman, Iran
e-mail: mrashidi@uk.ac.ir

M. MollahassaniPour
Faculty of Electrical and Computer Engineering, University of Sistan and Baluchestan, Zahedan, Iran
e-mail: m.mollahassani@ece.usb.ac.ir

 
V. Vahidinasab, B. Mohammadi-Ivatloo (eds.), *Demand-Side Peer-to-Peer Energy Trading*, Green Energy and Technology, https://doi.org/10.1007/978-3-031-35233-1_1

different solutions were suggested by international energy organizations all around the world to increase the hosting capacity of DERs in energy systems. The *energy community*, which was defined and regulated by European Union (EU), is one of the most interesting solutions to efficiently manage the smart distribution system with a high penetration rate of DERs. The energy community covers the most important programs such as Peer-to-Peer (P2P) energy trading, Energy Efficiency Programs (EEPs), and Demand Response Programs (DRPs), as well as the adoption of Energy Storage Systems (ESSs) and Electric Vehicles (EVs) in an integrated energy structure.

This chapter concentrates on the P2P energy trading concept as one of the most important features of energy communities where the role of energy communities in community energy is also investigated. Moreover, due to the critical role of people, as the heart of the energy system, in P2P energy trading, the effective behaviors of end users such as preferences, time discount impacts, and psychological time impacts are also studied. The control approaches of P2P energy trading, including centralized (or community-based), decentralized, and distributed, are also addressed. Furthermore, since the P2P energy trading brings some opportunities and challenges for end users in energy communities, the most crucial ups and downs are appraised. Finally, a vivid outlook of the current and future conditions of the P2P energy trading program is provided.

## 1.2 Community Energy and Energy Community Concepts

The energy transition from a conventional energy system into a decentralized, decarbonized, and democratized energy structure leads to more effective participation of end users in the energy systems. This energy transition contains different aspects, including economic, environmental, technical, and social. The economic, technical, and environmental features of the energy systems have been taken into consideration in energy policy researche where the social aspect is contemplated by developing the decentralized energy resources. Accordingly, comprehensive socio-political studies are required to evaluate the most critical aspects of socially equitable energy generation and energy distribution. Since the consumers play an active role in the future smart energy system, a new opportunity can be provided to integrate social science into energy studies. Thus, the *community energy* concept is nominated by appending the social aspects to the energy system structure. Community energy, as a complicated and comprehensive concept, covers all technical, social, environmental, and economic aspects of the energy systems. Different definitions are provided to describe the community in energy systems, as reported in [5] that 183 diverse definitions were presented. In community energy, the *energy* itself denotes various types of carriers such as electrical, thermal, gas, and water. Recently, the worldwide interest in the community energy concept has dramatically increased. A multi-purpose energy community was modeled in [6] to assess the effects of end users' attitudes on technical, social, environmental, and economic features of energy systems. It was shown that people involved in this community could be more aware of environmental issues and they could develop a more sustainable attitude. An agent-based modeling approach was used in [7] to model energy security in community energy systems. A bi-level chance-constrained programming optimization model of the energy management incorporating smart buildings in a community energy system was proposed in [8] which improved the operational flexibility as well as conviviance of both supply and demand sides.

On the other side, the international energy agencies and energy policymakers presented a new concept namely *Energy Community* which was considered as part of community energy concept. The energy community was defined as "*a group of consumers and/or prosumers, that together share energy generation units and electricity storage*" [5]. The European Union regularizes the energy communities by publishing the Clean Energy Package (CEP) in 2016 [9]. The CEP categorizes the energy community concept in two clusters: (i) Citizen Energy Communities (CECs) regarding to Directive (EU) 2019/944 (the recast Electricity Directive) and (ii) Renewable Energy Communities (RECs) regarding to Directive (EU) 2018/2001 (the recast renewable energy directive) [10]. The main target of CECs is facilitating energy generation, distribution, and storage, which covers all demand-side activities like DRPs, EEPs, P2P energy trading, and EVs. Moreover, CECs merely concentrate on electric energy section, whereas RECs cover all types of energy carriers, such as electrical, thermal, gas, and water. However, RECs merely support renewable energy resources whereas the CECs adopts all types of generation. RECs and CECs have been investigated from different standpoints in some researches. The challenges and advantages of CECs on the distribution side were evaluated in [11]. Under the CEC environment, the critical role of energy community participants' preferences on the household investment decisions over a long-time horizon was investigated in [12]. A market mechanism was proposed in [13] for local energy trading in the CEC structure. The optimal planning of RECs was addressed in [14] by proposing a mixed-integer linear programming optimization model. A linear optimization approach was developed in [15] to assess the self-consumption and self-sufficiency indices in a REC environment, where the proposed approach improved the operation status of the smart REC.

## 1.3 Transactive Energy Systems and P2P Energy Trading

The U.S. Department of Energy (DoE) defined transactive energy systems as "*a system of economic and control mechanisms that allows the dynamic balance of supply and demand across the entire electrical infrastructure using value as a key operational parameter*" [16]. Hence, transactive energy systems provide a market-based solution, implemented in different levels: (i) regional grid, (ii) local grid, (iii) distribution network, and (iv) energy communities, to efficiently coordinate generation and consumption in a smart environment. In a *regional grid*, the transactive energy system is a two-way energy and information exchange system between the local grids and the large-scale thermal power generation units which concentrates on improving interoperability, reliability, and efficiency [17]. In a *local grid*, transactive energy provides new services and opportunities for distribution networks [18]. In a *distribution network*, the transactive energy framework ameliorates the operational flexibility of the distribution system and the upstream network. In the *energy communities*, as the final layer of energy systems, the transactive energy system enables end users to trade energy among themselves and the distribution network under the supervision of the Energy Community Manager (ECM).

In the last few years, research studies have shown an increased interest in the transactive energy concept [1, 18] as well as its' applications [19, 20]. An analysis of current trends toward the transactive energy systems was provided in [18] aiming at defining the transactive energy system from different points of view and how it was coordinated in regard to the smart grid environment. A systematic literature review of local energy market models considering transactive energy system, P2P energy trading, and community self-consumption was presented in [1]. In [19], a bi-level strategic energy trading structure was proposed to minimize the operation cost of the distribution network, considering transactive energy hubs and EVs. A bi-level transactive model for the networked microgrids was presented in [20] aiming at improving operation efficiency, reliability, and flexibility of the distribution network and microgrids.

Here, a sample schematic of the distribution network of a sustainable city associated with two energy communities[1] is portrayed in Fig. 1.1. The main concentration of Fig. 1.1 is on the concept of transactive energy systems where there is much of interest to specifically assess the role of energy communities and P2P energy trading in distribution systems. As shown in Fig. 1.1, prosumers can either buy/sell the extra/slack electricity from/to the Local Energy Market (LEM) in each energy community or can trade energy with the upstream distribution network based upon the Feed-in Tariff (FIT) price which is determined by the Distribution System Operator (DSO). Correspondingly, consumers have an opportunity to select the supplied-energy resource, either through LEM or upstream distribution network. Furthermore, as can be seen in Fig. 1.1, different types of DERs including wind turbines, microturbines, PV systems, and energy storages are appended to the main body of the distribution system which are owned by private individuals. Hence, DSO can efficiently interchange energy with ECMs and transmission network to optimally balance the demand and supply in the distribution system. Finally, it is worth mentioning that other energy carriers such as thermal, gas, and water can be also considered in the energy communities.

[1]It is obvious that Fig. 1.1 provides a typical distribution system and the number of energy communitis in the real distribution system can be more.

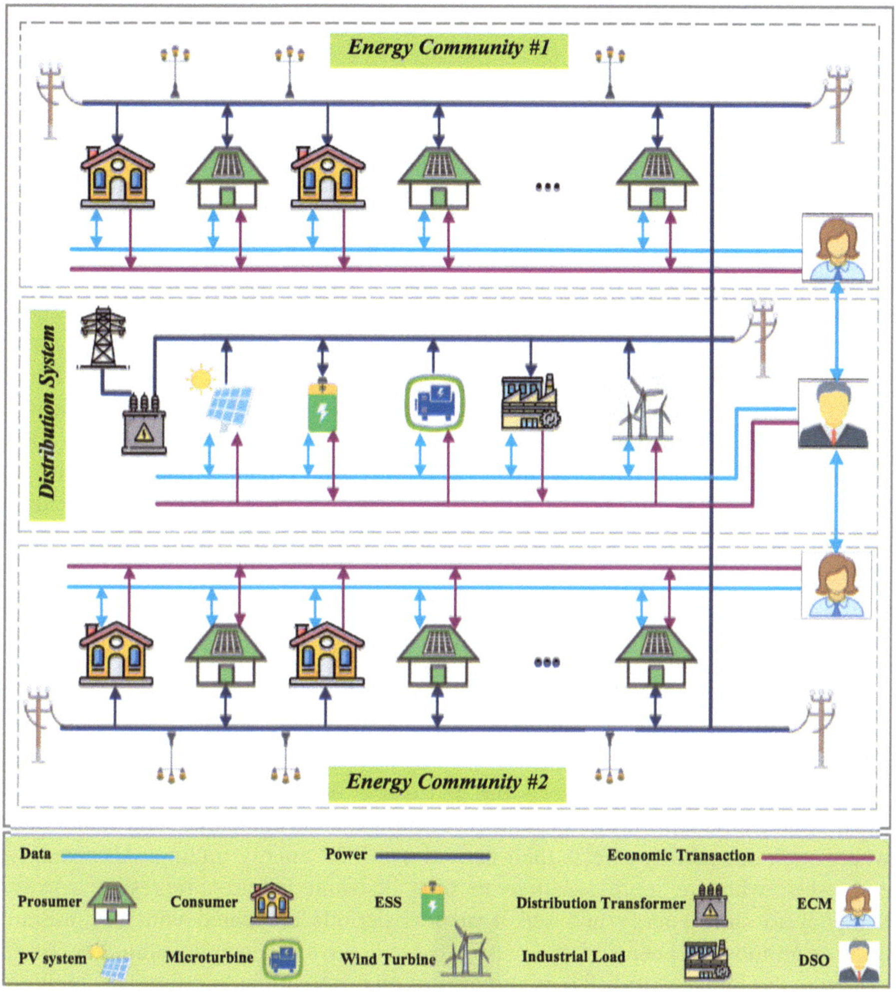


**Fig. 1.1** The structure of a sustainable distribution system associated with energy communities

## 1.4 Role of End Users' Behavior in P2P Energy Trading

The end users' behaviors are contemplated as the most crucial factor to determine the level of exchanged energy and P2P energy trading's prices since the individuals are the main decision maker in the P2P energy trading framework. The utility function of each end user is distinct and reliant on their individual attitudes. Hence, the most effective behavioral parameters of end users are elucidated in detail in the following:

### 1.4.1 Preferences

Under traditional energy systems, electrical energy is considered a homogeneous product, where the preferences of end users were not competently contemplated in the design and operation of the energy systems. On the contrary, under the P2P energy trading framework, end users can actively play diverse roles in all sections (generation, storage, and consumption) since such an environment offers an opportunity to appraise the heterogeneous preferences of individuals. As an example, some end users show environmental preferences in local P2P energy trading, which represents their willingness to pay more to use generated power of green energy resources compared with conventional non-renewable resources. Therefore, novel technologies should be utilized in the energy communities to satisfy the environmental preferences of end users. As real cases, the end users in Piclo and Vandebron online platforms, which are P2P energy trading platforms, have the ability to track and select the type of energy resources [21]. Moreover, philanthropic behaviors can be also regarded as one of the common preferences in the P2P energy trading problem. In this regard, some end users have a desire to share the extra generated energy with low-income households which can be interpreted as philanthropic preferences.

### 1.4.2 Time Discount Impacts

The time discount concept affects the end users' behaviors from a psychological standpoint. Individuals prefer present economic benefits compared with future ones where such desirability affects their activities in P2P energy trading. Accordingly, end users, who are more sensitive to time discount impacts, represent a lower tendency to store energy and sell it in next periods compared with the present. Such an attitude has been widely investigated in behavioral economics studies. As an example, the impacts of different attitudes of end users including time discounting, risk aversion, loss aversion, and present bias on household adoption of energy-efficient technologies were evaluated in [22] based on a large-scale empirical test. It was concluded that end users with a higher time discount attitude had a lower tendency to invest in high-efficiency devices.

### 1.4.3 Psychological Time Impacts

The psychological time is a product of an individual's mind which can affect his/her behaviors. Accordingly, the individual's utility function can alter based on his/her brain's understanding of encountering physical time. Hence, an understanding of a specified physical time duration is entirely different for people due to the psychology of time perception.

Investigation of the impacts of psychological time on future smart energy systems studies seems imperative due to the effective role of end users in such an environment. As an example, psychological time impacts were modeled in the home energy management problem [23]. It was shown that psychological time consideration affected the end users' behaviors in the use of different home appliances which altered the optimal scheduling of energy management problem.

### *1.4.4 Reference Point Impacts*

The mental reference point determines how an outcome of a strategy is perceived in someone's mind, which is different from person to person. To make a decision, an individual compares the possible result with a mental reference point depending on his/her past experiences and future aspirations.

The psychological impacts of reference point have been investigated in some energy systems studies. The favorable indoor temperature as the end user reference point in the home energy management problem was modeled in [24]. The impacts of diverse monetary reference points on end-user satisfaction in the home energy management problem was evaluated in [23]. The results showed that end-user satisfaction diminished by increasing the level of monetary reference point. The impacts of mental reference point on the status of the P2P energy trading was appraised in [25] where it was shown that the grid energy exchanged in the energy community was significantly dependent on the prosumers' reference points.

### *1.4.5 Loss Aversion Impacts*

Loss aversion as a cognitive bias is a tendency in behavioral economics studies. This feature describes that a real or potential loss is recognized as more severe compared with an equivalent gain from a psychological standpoint. Hence, individuals are emotionally affected more by the loss of money or other precious objects than by making gains.

The impacts of loss aversion feature have been scrutinized in future energy systems studies. The role of risk preferences and loss aversion in the use of energy-efficient appliances was investigated in [26]. It was concluded that the person with more profound loss aversion attitude possessed more willingness to use durable energy-efficient appliances, where the demographic factors such as age, gender, education, and family location had significant effects on appliance use behaviors. The impact of loss aversion in the community-based P2P energy trading problem was investigated in [27] due to the significant role of end users' utility function in such an environment. It was shown that the variation of personal behaviors affected the payment cost of energy community.

### *1.4.6 Risk-Averse and Risk-Taker Behaviors*

Humans encounter different types of uncertainty in all decision-making situations which are handled based on the utility function of his/her mind in such a condition [28]. From a psychological point of view, the end user utility function is divided into two sections: (i) gain and (ii) loss based on reference point concept [29], where the human behavior are not identical in these two areas [30]. The experimental studies in behavioral economic show that people are risk-averse/takers in gain/loss area of the utility function, respectively [31].

In an energy system, the end users face diverse uncertainties such as energy demand, upstream energy price, and generation of renewable resources. Therefore, it is essential to address the challenges associated with modeling risk-averse and risk-taking behavior of end-users in energy scheduling. As an example, the risk-averse/taker behaviors of end users were modeled by the prospect theory approach [27] and the conditional value at risk technique [32] in P2P energy trading problem.

## 1.5 Energy Trading Market Structures

The energy trading is handled by different control approaches in an energy community where each one possesses some strengths and weaknesses. Here, the different control structures of the local energy trading market, including (i) centralized or community-based, (ii) distributed, and (iii) decentralized, are entirely elaborated. In summary, Table 1.1 indicate the comparison of the various market structures.

### *1.5.1 Centralized*

In a centralized approach, also named community-based approach, the ECM plays a fundamental role in P2P energy trading market structure. Referring to the CEP, the ECM is a non-profit entity that can be either a local data center system or a human who coordinates sellers and buyers' energy transaction and clears the local energy market. The ECM acts as a medium between the upper-level energy market and the energy community's participants which keeps the small-scale participants from the high price variations at the upper-level. Moreover, the ECM considers the power flow constraints in the optimization problem to prevent congestions in the network.

**Table 1.1** Market structures comparison

| | Privacy | Convergence speed | Solution optimality | Reliability |
|---|---|---|---|---|
| Centralized | Low | High | High | Low |
| Distributed | High | Low | Low | High |
| Decentralized | Medium | Medium | Medium | Medium |

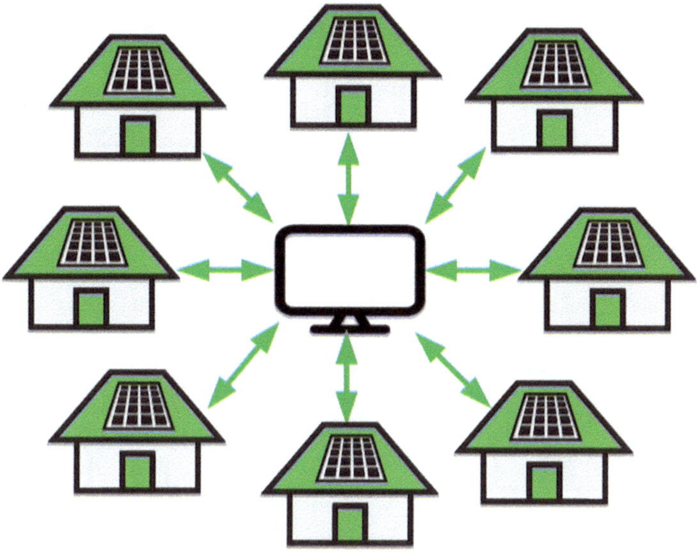

**Fig. 1.2** The community-based energy trading structure

Although the centralized approach provides the most optimal solutions compared with other ones, it brings some critical challenges regarding the end users' privacy and scalability. The schematic of the community-based control structure is depicted in Fig. 1.2.

The community-based energy trading structure is addressed in some novel researches. An overview of the potential roles of the collective energy community and blockchain to go beyond P2P energy trading was provided in [33]. A local energy community model considering P2P energy trading and DRPs was proposed in [34] to reduce the energy bills of the smart community. An efficient bidding-based P2P energy trading model considering the green energy preference in the virtual energy community was presented in [35].

### *1.5.2 Distributed*

In a distribution approach, each participant can directly negotiate with other contributors through a highly accurate and complicated communication infrastructure in the energy community without the intervention of a central entity. Hence, the distributed approach provides more privacy for end users and less complexity in the implementation stage compared with the centralized method. However, scalability is still a big challenge in the market design stage since the number of potential negotiations among participants will increase in large-scale energy communities, which leads to a slow convergence rate. Additionally, considering the power flow

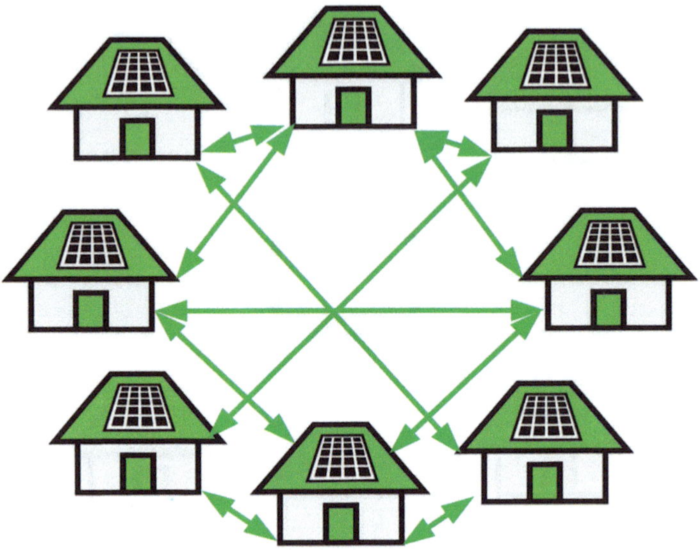

**Fig. 1.3** The distributed energy trading structure

constraints in the presence of different energy transaction options in the distributed energy trading structure is also a challenging issue. A sample schematic of the distributed energy trading structure is shown in Fig. 1.3.

This control strategy has been utilized to effectively manange the energy systems. As an example, a distributed P2P energy trading mechanism was proposed in [36] to manage energy trading among end users, where this control approach was addressed in [37] with consideration of prosumers' subjective preferences and social behaviors.

### *1.5.3 Decentralized*

The decentralized approach is organized as a combinatory structure of centralized and distributed approaches. Here, the participants can contribute to energy trading in an energy community regarding the ECM control decisions, where energy and data information is shared among other energy communities to guarantee a more reliable, cost-efficient, and flexible operation. This approach has a low convergence time compared with the distributed structure which brings higher scalability compared with a centralized approach due to the presence of more ECMs. However, the optimal solution obtained in the decentralized approach is weaker than the centralized structure and, the end user's privacy is placed slightly in a lower degree. A sample schematic of the decentralized energy trading structure is portrayed in Fig. 1.4.

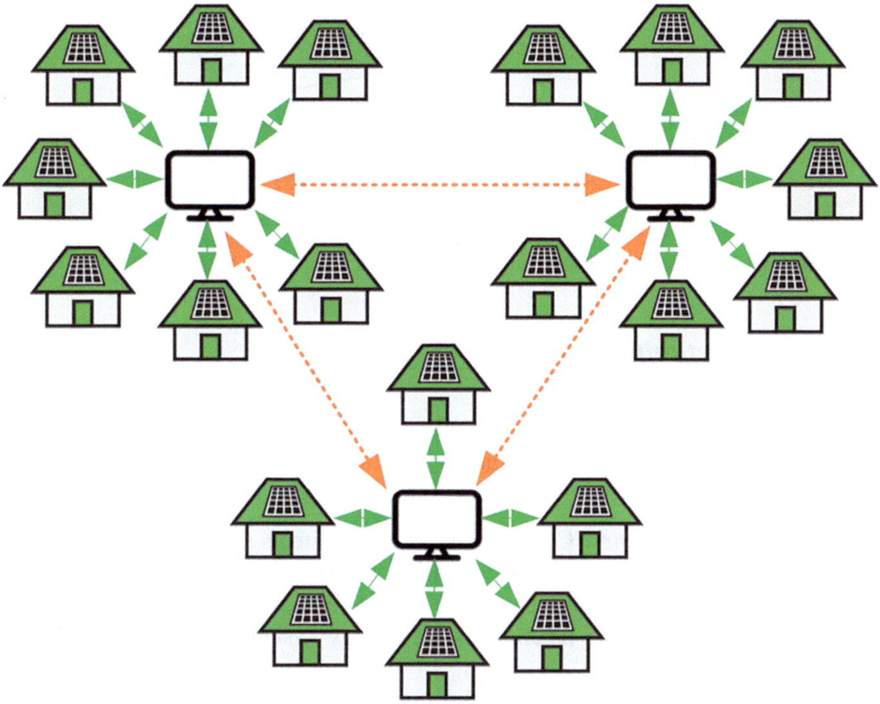

**Fig. 1.4** The decentralized energy trading structure

The decentralized approach is appraised in some research related to P2P energy trading problem. In [38] a blockchain-based decentralized hybrid P2P energy trading model was provided. A security-constrained decentralized P2P energy trading in a smart energy system was addressed in [39]. An optimal decentralized market clearing platform for P2P energy trading in a grid energy system was presented in [40].

## 1.6 Challenges and Opportunities of P2P Energy Trading

Nowadays, increasing the penetration rate of DERs in the demand side of energy systems motivates researchers to develop the P2P energy trading models [41, 42]. This remodeling brings some challenges and opportunities to the smart energy systems. The challenges and opportunities of P2P energy trading are addressed in Fig. 1.5. In the following, the main benefits and difficulties of P2P energy trading implementation from technical, economic, environmental, and social points of view in the local energy markets are elucidated.

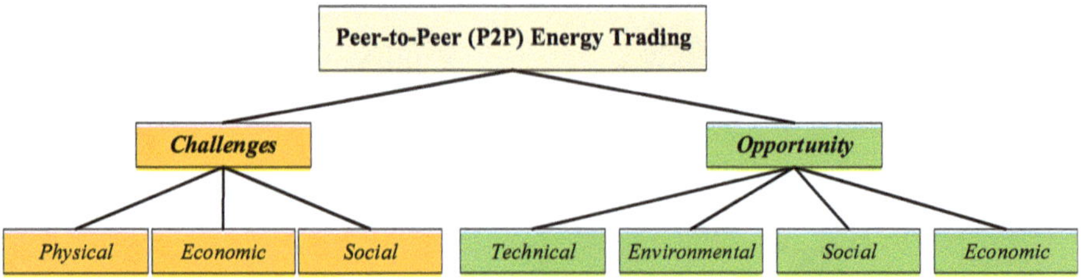


**Fig. 1.5** The challenges and opportunities of P2P energy trading

## 1.6.1 Opportunities

The adoption of P2P energy trading will change the role of individual consumers from passive participants to active ones in the energy community. This issue brings significant advantages to energy system users, which are discussed in the following:

- *Technical opportunities:* Conventional energy systems suffer from critical challenges such as power losses and transmission line congestion [43]. Local energy trading in energy communities can be considered a promising solution to overcome difficulties which ameliorates the energy system performance from the technical standpoint.
- *Economic opportunities:* Individual economic benefits have been identified as a motivation for humans to change their performance. As an example, reducing energy bills inspires consumers to utilize personal DERs like PV systems, who are named prosumers. Such an economic motivation encourages the prosumers to join energy communities [44] and actively participate in P2P energy trading. Hence, the consumers' demand can be satisfied in the local P2P energy trading market at a lower price compared to the upstream network. Moreover, the prosumers also possess the opportunity to sell the excess generated energy of DERs at a higher price under the P2P energy trading structure compared with the FIT price of the upstream network. Accordingly, such circumstances provide a win-win game for both consumers and prosumers in energy communities [45] which is assessed in different studies. As an example, it was shown in [46] that P2P energy trading based on human decision-making led to financial benefits for energy community participants and reduced stress for the upstream network. The economic benefits of participation in local energy markets were observed in [47] due to the use of green energy resources in a renewable-based energy community.
- *Environmental opportunities:* The P2P energy trading structure can provide environmental benefits in the energy communities due to the possibility of using local resources, especially green resources [48, 49]. The outcomes of the real case in the Netherlands showed that the environmental benefits possessed extreme value in peer-to-peer electricity trading [50]. Moreover, it was reported that environmental concerns were contemplated as a motivation in joining prosumers to the collective self-consumption energy communities [44]. Findings

of a study on German community-based renewable energy initiatives showed that environmental motivations had a more profound role compared with financial return in P2P energy trading [51].

- *Social opportunities:* Social factors have a drastic impact on inspiring end users to become prosumers. Hence, the participation of prosumers in P2P energy trading positively affects the social norms in an energy community. The transition role of end users from passive participants in conventional energy systems to active ones in energy communities provides an opportunity to select the energy resources. Accordingly, it is expected that the energy democracy indices will improve with the advent of energy communities.

### *1.6.2 Challenges*

The main challenges of the P2P energy trading problem are as follows:

- *Physical challenges:* The local distribution network faces some challenges in the presence of local energy trading like voltage variations (fluctuations), power imbalance, and distribution line congestion [52]. The main driver of voltage variations in energy communities arises from the numerous simultaneous P2P energy transactions among end users in the LEM as a result of the high penetration rate of DERs. However, the voltage fluctuations may negatively affect the function of inverters which lead to curtailing some P2P energy transactions. It was shown in [51] that the level of elasticities and prices could efficiently affect the voltage profile and, accordingly, could improve the performing status of P2P energy trading among end users. In [53], the issue of voltage fluctuations in low-voltage distribution feeder in presence of P2P energy trading was investigated and some preventive actions were presented to heal such difficulties. Moreover, the high penetration rate of P2P energy transactions can violate the power phase balance in the distribution system which is appraised in some studies. As an example, an efficient approach for increasing the hosting capacity of the distribution system through local energy markets and dynamic phase switching was proposed in [54]. Additionally, the impacts of P2P energy trading on increasing line congestion in the local distribution network seemed considerable [52]. However, utilizing energy storage systems [55, 56], and demand response programs [57] can relieve congestion problems in the local distribution power system.
- *Economic challenges:* Despite economic opportunities, significant economic challenges can be brought to system users in energy systems due to the implementation of P2P energy transactions. Such transactions among energy community participants provide profit for both consumers and prosumers [58], as clarified in the previous section. However, the level of purchased energy, from either the upstream energy retailer or the distribution network, is declined, which definitely leads to diminishing their revenues with the development of the energy

communities. To overcome such financial losses, the upstream retailer and the distribution network increase the price of electricity which merely imposes financial burdens on the single poor consumers who are out of the energy communities'territory [59].

- *Social challenges:* The P2P energy trading enables complicated technologies on the demand side of the energy system. Under such an environment, the end users also possess an innovative opportunity to satisfy demand from the LEM in energy communities. However, this promising opportunity brings two significant challenges from end users and the system operator points of view as follows:

  (i) The P2P energy trading may exacerbate existing inequalities for disengaged end users. A novel published policy brief paper on the *nature energy journal* claimed that: "*Innovative energy business models, such as peer-to-peer trading or energy as a service, are attractive to different groups of customers. Disengaged consumers with low trust in the energy market could face further disadvantages, while already active consumers could reap even greater benefits, which risks widening existing socio-economic inequalities*" [60].

  (ii) Commonly, technical- and economic-driven measures are the most crucial aspects in all energy policy decisions of energy systems. With the advent of energy communities and P2P energy trading concepts, the outlook of energy system studies has been altered and, accordingly, both behavioral and social features are also taken into consideration due to the active role of end users in energy communities [61]. Hence, adding these new features makes the energy system problems more sophisticated from a system operator standpoint.

## 1.7 P2P Energy Trading Outlook and Real Projects

The energy system outlook will become fabulous and promising with the development of energy communities in smart sustainable cities. The scientific studies clarify that energy communities provide a more green and habitable environment for humanity [62]. The energy communities can promote the novel programs and technologies such as P2P energy trading, EEPs, and DRPs, as well as the adoption of ESSs and EVs. However, it should be noted that P2P energy trading as a promising solution is the main aspect of energy communities.

In the last decades, P2P energy trading has been taken into contemplation in a considerable number of countries. The European continent, especially Germany, UK, Spain, and the Netherlands are the leading countries in P2P energy trading from both the R & D projects and real-world trials points of view [63]. A cloud-based P2P energy trading platform has been implemented in Germany for energy trading among smart homes [64]. In the UK, the famous P2P energy trading project, the so-called Piclo, is a software framework for energy and flexibility services trading between peers [65]. A blockchain-based market platform called Powerpeers has been used in the residential building section of the Netherlands to facilitate energy sharing between end users [66]. In North America, P2P energy trading projects have been well developed. As an example, a P2P energy trading project has been performed in the Brooklyn energy community testbed, where the participants in the microgrid can locally share their generated energy using the distribution system [67]. Moreover, the prominence of P2P energy trading has recently drawn great attention in Australia due to the extensive government subsidy from both the Federal and State governments [68]. The RENew Nexus project in Australia has provided an efficient opportunity for households to trade excess generated energy of rooftop PVs with other participants [69] while concentrating on evaluating the potential and challenges of localized energy markets. The National Energy Market (NEM) in Australia has developed a commercial P2P energy trading project to increase the use of green energy resources. In addition, a few constructive efforts have been made about P2P energy trading in the Asian territory. The Australia-based Power Ledger and the Japanese solar provider have actively striven to provide a testbed energy-sharing platform in Kanto, Japan [70]. A blockchain-based platform for P2P energy trading has been demonstrated in the Kanto region of Japan [71]. In India's Lucknow, a pilot project of P2P energy trading has been implemented for energy transactions among PV systems and end users [72]. Similarly, some advanced P2P energy trading projects are trialed in Thailand, Malaysia, and South Korea [63].

## 1.8 Prosumers' Participation in P2P Energy Trading

Prosumer participation rate in P2P energy trading has a significant impact on the productivity of the energy community. In this regard, finding an appropriate approach to motivate the prosumers is of much interest. In fact, social characteristics have a vital role in motivating end users to become prosumers and participate in P2P energy trading. Specifically, some studies have indicated the impact of peer effects and social influences on the adoption of PV systems [73]. Moreover, [74] claims that prosumers' personal relationship with their peer or community has a strong impact on the amount of prosumer energy sharing. On the other hand, the use of high-technology devices can be useful to facilitate P2P energy trading between prosumers. In this regard, proposing a fast and reliable P2P trading platform is an appropriate solution to enable automatic transactions which require minimal effort.

## 1.9 Summary

In this chapter, an overview of P2P energy trading framework as the main driver to obtaining a sustainable energy community was provided. In this regard, the concept of community energy was comprehensively elucidated and the role of the energy communities as the heart of the community energy was addressed. Moreover, the most effective behavioral attitudes of end users in P2P energy trading problem were evaluated whereas the main opportunities and challenges of such framework were scrutinized. Finally, an outlook of the current and future conditions of the P2P energy trading program was appraised. The main achievements of this chapter can be summarized as follows:

- The P2P energy trading in energy communities enables people to play a vital role in the scheduling of energy systems. Hence, the significance of social aspects is considerably increased in energy systems.
- Individual attitudes such as preferences, time discount, psychological time, reference point, loss aversion, risk-averse, and risk-taker behaviors are the most effective parameters in the end users' utility function.
- The energy community structure can be managed by various control topologies such as centralized, decentralized, and distributed. However, the decentralized control approach is more prevalent in the P2P energy trading optimization models.
- The P2P energy trading provides advantages for the energy community participants and the upstream energy system. The environmental benefits of P2P energy trading implementation are the most important momentum for participants to become an active prosumer. However, some physical, economic, and social challenges can flog the development path of P2P energy trading in the world.
- The investigation of real-world projects shows that a profound intention is existing to develop P2P energy trading platforms in the world. Therefore, the future of P2P energy trading in the energy community can be promising compared with current situation.

## References

1. Capper, T., et al. (2022, July). Peer-to-peer, community self-consumption, and transactive energy: A systematic literature review of local energy market models. *Renewable and Sustainable Energy Reviews, 162*, 112403. https://doi.org/10.1016/j.rser.2022.112403
2. Gatto, A. (2022, July). The energy futures we want: A research and policy agenda for energy transitions. *Energy Research and Social Science, 89*, 102639. https://doi.org/10.1016/j.erss.2022.102639
3. Tushar, W., Saha, T. K., Yuen, C., Smith, D., & Poor, H. V. (2020, July). Peer-to-peer trading in electricity networks: An overview. *IEEE Transactions on Smart Grid, 11*(4), 3185–3200. https://doi.org/10.1109/TSG.2020.2969657
4. Dorahaki, S., Dashti, R., & Shaker, H. R. (2020, October). Optimal outage management model considering emergency demand response programs for a smart distribution system. *Applied Sciences, 10*(21), 7406. https://doi.org/10.3390/app10217406
5. Bauwens, T., et al. (2022, March). Conceptualizing community in energy systems: A systematic review of 183 definitions. *Renewable and Sustainable Energy Reviews, 156*, 111999. https://doi.org/10.1016/j.rser.2021.111999
6. Ceglia, F., Esposito, P., Faraudello, A., Marrasso, E., Rossi, P., & Sasso, M. (2022, October). An energy, environmental, management and economic analysis of energy efficient system

towards renewable energy community: The case study of multi-purpose energy community. *Journal of Cleaner Production, 369*, 133269. https://doi.org/10.1016/j.jclepro.2022.133269
7. Fouladvand, J., Ghorbani, A., Sarı, Y., Hoppe, T., Kunneke, R., & Herder, P. (2022, September). Energy security in community energy systems: An agent-based modelling approach. *Journal of Cleaner Production, 366*, 132765. https://doi.org/10.1016/j.jclepro.2022.132765
8. Su, S., Li, Z., Jin, X., Yamashita, K., Xia, M., & Chen, Q. (2022, June). Bi-level energy management and pricing for community energy retailer incorporating smart buildings based on chance-constrained programming. *International Journal of Electrical Power & Energy Systems, 138*, 107894. https://doi.org/10.1016/j.ijepes.2021.107894
9. Lowitzsch, J., Hoicka, C. E., & van Tulder, F. J. (2020, April). Renewable energy communities under the 2019 European Clean Energy Package – Governance model for the energy clusters of the future? *Renewable and Sustainable Energy Reviews, 122*, 109489. https://doi.org/10.1016/j.rser.2019.109489
10. Capros, P., et al. (2018, November). Outlook of the EU energy system up to 2050: The case of scenarios prepared for European Commission's 'clean energy for all Europeans' package using the PRIMES model. *Energy Strategy Reviews, 22*, 255–263. https://doi.org/10.1016/j.esr.2018.06.009
11. Stroink, A., Diestelmeier, L., Hurink, J. L., & Wawer, T. (2022, March). Benefits of cross-border citizen energy communities at distribution system level. *Energy Strategy Reviews, 40*, 100821. https://doi.org/10.1016/j.esr.2022.100821
12. Golla, A., Röhrig, N., Staudt, P., & Weinhardt, C. (2022, August). Evaluating the impact of regulation on the path of electrification in Citizen Energy Communities with prosumer investment. *Applied Energy, 319*, 119241. https://doi.org/10.1016/j.apenergy.2022.119241
13. Oprea, S.-V., Bâra, A., & Diaconita, V. (2022, June). A motivational local trading framework with 2-round auctioning and settlement rules embedded in smart contracts for a small citizen energy community. *Renewable Energy, 193*, 225–239. https://doi.org/10.1016/j.renene.2022.04.149
14. Cosic, A., Stadler, M., Mansoor, M., & Zellinger, M. (2021, December). Mixed-integer linear programming based optimization strategies for renewable energy communities. *Energy, 237*, 121559. https://doi.org/10.1016/j.energy.2021.121559
15. Cielo, A., Margiaria, P., Lazzeroni, P., Mariuzzo, I., & Repetto, M. (2021, September). Renewable Energy Communities business models under the 2020 Italian regulation. *Journal of Cleaner Production, 316*, 128217. https://doi.org/10.1016/j.jclepro.2021.128217
16. The GridWise Architecture Council GridWise Transactive Energy Framework Version 1.0. (2015). https://www.gridwiseac.org/pdfs/te_framework_report_pnnl-22946.pdf. Accessed 01 Sept 2022.
17. Knight, M. R. (2016, December). Stochastic impacts of metaheuristics from the toaster to the turbine [technology leaders]. *EEE Electrification Magazine, 4*(4), 52–40. https://doi.org/10.1109/MELE.2016.2614218
18. Abrishambaf, O., Lezama, F., Faria, P., & Vale, Z. (2019, November). Towards transactive energy systems: An analysis on current trends. *Energy Strategy Reviews, 26*, 100418. https://doi.org/10.1016/j.esr.2019.100418
19. Najafi, A., Pourakbari-Kasmaei, M., Jasinski, M., Contreras, J., Lehtonen, M., & Leonowicz, Z. (2022, August). The role of EV based peer-to-peer transactive energy hubs in distribution network optimization. *Applied Energy, 319*, 119267. https://doi.org/10.1016/j.apenergy.2022.119267
20. Wu, Y., Shi, J., Lim, G. J., Fan, L., & Molavi, A. (2020, November). Optimal management of transactive distribution electricity markets with co-optimized bidirectional energy and ancillary service exchanges. *IEEE Transactions on Smart Grid, 11*(6), 4650–4661. https://doi.org/10.1109/TSG.2020.3003244

21. Morstyn, T., Farrell, N., Darby, S. J., & McCulloch, M. D. (2018, February). Using peer-to-peer energy-trading platforms to incentivize prosumers to form federated power plants. *Nature Energy, 3*(2), 94–101. https://doi.org/10.1038/s41560-017-0075-y
22. Schleich, J., Gassmann, X., Meissner, T., & Faure, C. (2019, May). A large-scale test of the effects of time discounting, risk aversion, loss aversion, and present bias on household adoption of energy-efficient technologies. *Energy Economics, 80*, 377–393. https://doi.org/10.1016/j.eneco.2018.12.018
23. Dorahaki, S., Rashidinejad, M., Fatemi Ardestani, S. F., Abdollahi, A., & Salehizadeh, M. R. (2022, April). A home energy management model considering energy storage and smart flexible appliances: A modified time-driven prospect theory approach. *Journal of Energy Storage, 48*, 104049. https://doi.org/10.1016/j.est.2022.104049
24. Good, N. (2019, April). Using behavioural economic theory in modelling of demand response. *Applied Energy, 239*, 107–116. https://doi.org/10.1016/j.apenergy.2019.01.158
25. El Rahi, G., Etesami, S. R., Saad, W., Mandayam, N. B., & Poor, H. V. (2019, January). Managing price uncertainty in prosumer-centric energy trading: A prospect-theoretic Stackelberg game approach. *IEEE Transactions on Smart Grid, 10*(1), 702–713. https://doi.org/10.1109/TSG.2017.2750706
26. He, R., Jin, J., Gong, H., & Tian, Y. (2019, April). The role of risk preferences and loss aversion in farmers' energy-efficient appliance use behavior. *Journal of Cleaner Production, 215*, 305–314. https://doi.org/10.1016/j.jclepro.2019.01.076
27. Dorahaki, S., Rashidinejad, M., Ardestani, S. F. F., Abdollahi, A., & Salehizadeh, M. R. (2021, September). A peer-to-peer energy trading market model based on time-driven prospect theory in a smart and sustainable energy community. *Sustainable Energy, Grids and Networks, 28*, 100542. https://doi.org/10.1016/J.SEGAN.2021.100542
28. Ziras, C., Sousa, T., & Pinson, P. (2021, September). What do prosumer marginal utility functions look like? Derivation and analysis. *IEEE Transactions on Power Apparatus and Systems, 36*(5), 4322–4330. https://doi.org/10.1109/TPWRS.2021.3068620
29. Jhala, K., Natarajan, B., & Pahwa, A. (2019, May). Prospect theory-based active consumer behavior under variable electricity pricing. *IEEE Transactions on Smart Grid, 10*(3), 2809–2819. https://doi.org/10.1109/TSG.2018.2810819
30. Frederiks, E. R., Stenner, K., & Hobman, E. V. (2015, January). Household energy use: Applying behavioural economics to understand consumer decision-making and behaviour. *Renewable and Sustainable Energy Reviews, 41*, 1385–1394. https://doi.org/10.1016/J.RSER.2014.09.026
31. Kahneman, D., & Tversky, A. (2013). Prospect theory: An analysis of decision under risk. *Econometrica, 47*(2), 99–127.
32. Lin, W.-T., Chen, G., & Li, C. (2021, November). Risk-averse energy trading among peer-to-peer based virtual power plants: A stochastic game approach. *International Journal of Electrical Power & Energy Systems, 132*, 107145. https://doi.org/10.1016/j.ijepes.2021.107145
33. Wu, Y., Wu, Y., Cimen, H., Vasquez, J. C., & Guerrero, J. M. (2022, May). Towards collective energy community: Potential roles of microgrid and blockchain to go beyond P2P energy trading. *Applied Energy, 314*, 119003. https://doi.org/10.1016/j.apenergy.2022.119003
34. Zhou, S., Zou, F., Wu, Z., Gu, W., Hong, Q., & Booth, C. (2020, January). A smart community energy management scheme considering user dominated demand side response and P2P trading. *International Journal of Electrical Power & Energy Systems, 114*, 105378. https://doi.org/10.1016/j.ijepes.2019.105378
35. Park, D.-H., Park, J.-B., Lee, K. Y., Son, S.-Y., & Roh, J. H. (2021). A bidding-based peer-to-peer energy transaction model considering the green energy preference in virtual energy community. *IEEE Access, 9*, 87410–87419. https://doi.org/10.1109/ACCESS.2021.3061767
36. Yao, Y., Yang, J., Chen, S., Gao, C., & Chen, T. (2020, September). Design of distributed power trading mechanism based on P2P contract. In *2020 12th IEEE PES Asia-Pacific power and energy engineering conference (APPEEC)* (pp. 1–5). https://doi.org/10.1109/APPEEC48164.2020.9220421

37. Yao, Y., Gao, C., Chen, T., Yang, J., & Chen, S. (2021, October). Distributed electric energy trading model and strategy analysis based on prospect theory. *International Journal of Electrical Power & Energy Systems, 131*, 106865. https://doi.org/10.1016/j.ijepes.2021.106865
38. Solanki, B., et al. (2021, December). Blockchain-based decentralized hybrid P2P energy trading. In *2021 9th IEEE international conference on power systems (ICPS)* (pp. 1–5). https://doi.org/10.1109/ICPS52420.2021.9670159
39. Wang, L., et al. (2022). Security constrained decentralized peer-to-peer transactive energy trading in distribution systems. *CSEE Journal of Power and Energy Systems, 8*(1), 188–197. https://doi.org/10.17775/CSEEJPES.2020.06560
40. Hung, N. M., & Ahn, H.-S. (2021). An optimal market clearing algorithm for peer-to-peer energy trading in smart grid. In *2021 21st international conference on control, automation and systems (ICCAS)* (pp. 1071–1075). https://doi.org/10.23919/ICCAS52745.2021.9649818
41. Dorahaki, S., Rashidinejad, M., Ardestani, S. F. F., Abdollahi, A., & Salehizadeh, M. R. (2023, January). Probabilistic/information gap decision theory-based bilevel optimal management for multi-carrier network by aggregating energy communities. *IET Renewable Power Generation, 17*, 1436. https://doi.org/10.1049/rpg2.12685
42. Dorahaki, S., Sarkhosh, A., Rashidinejad, M., Salehizadeh, M. R., & MollahassaniPour, M. (2023, March). Fairness in optimal operation of transactive smart networked modern multi-carrier energy systems: A two-stage optimization approach. *Sustainable Energy Technologies and Assessments, 56*, 103035. https://doi.org/10.1016/j.seta.2023.103035
43. Rahman, S., et al. (2022, July). A framework to assess voltage stability of power grids with high penetration of solar PV systems. *International Journal of Electrical Power & Energy Systems, 139*, 107815. https://doi.org/10.1016/j.ijepes.2021.107815
44. Bauwens, T. (2019, June). Analyzing the determinants of the size of investments by community renewable energy members: Findings and policy implications from Flanders. *Energy Policy, 129*, 841–852. https://doi.org/10.1016/j.enpol.2019.02.067
45. Lüth, A., Zepter, J. M., Crespo del Granado, P., & Egging, R. (2018, November). Local electricity market designs for peer-to-peer trading: The role of battery flexibility. *Applied Energy, 229*, 1233–1243. https://doi.org/10.1016/j.apenergy.2018.08.004
46. Pena-Bello, A., Parra, D., Herberz, M., Tiefenbeck, V., Patel, M. K., & Hahnel, U. J. J. (2022, January). Integration of prosumer peer-to-peer trading decisions into energy community modelling. *Nature Energy, 7*(1), 74–82. https://doi.org/10.1038/s41560-021-00950-2
47. Jogunola, O., et al. (2017, December). State-of-the-art and prospects for peer-to-peer transaction-based energy system. *Energies, 10*(12), 2106. https://doi.org/10.3390/en10122106
48. Wittenberg, I., & Matthies, E. (2016, November). Solar policy and practice in Germany: How do residential households with solar panels use electricity? *Energy Research and Social Science, 21*, 199–211. https://doi.org/10.1016/j.erss.2016.07.008
49. Palm, J. (2018, February). Household installation of solar panels – Motives and barriers in a 10-year perspective. *Energy Policy, 113*, 1–8. https://doi.org/10.1016/j.enpol.2017.10.047
50. Georgarakis, E., Bauwens, T., Pronk, A.-M., & AlSkaif, T. (2021, December). Keep it green, simple and socially fair: A choice experiment on prosumers' preferences for peer-to-peer electricity trading in the Netherlands. *Energy Policy, 159*, 112615. https://doi.org/10.1016/j.enpol.2021.112615
51. Radtke, J. (2014, December). A closer look inside collaborative action: Civic engagement and participation in community energy initiatives. *People, Place and Policy Online, 8*(3), 235–248. https://doi.org/10.3351/ppp.0008.0003.0008
52. Dudjak, V., et al. (2021, November). Impact of local energy markets integration in power systems layer: A comprehensive review. *Applied Energy, 301*, 117434. https://doi.org/10.1016/j.apenergy.2021.117434
53. Azim, M. I., Pourmousavi, S. A., Tushar, W., & Saha, T. K. (2019, August). Feasibility study of financial P2P energy trading in a grid-tied power network. In *2019 IEEE Power & Energy Society general meeting (PESGM)* (pp. 1–5). https://doi.org/10.1109/PESGM40551.2019.8973809

54. Horta, J., Kofman, D., Menga, D., & Caujolle, M. (2018, June). Augmenting DER hosting capacity of distribution grids through local energy markets and dynamic phase switching. In *Proceedings of the ninth international conference on future energy systems* (pp. 314–318). https://doi.org/10.1145/3208903.3208937
55. Santos, J. M., Moura, P. S., & de Almeida, A. T. (2014, September). Technical and economic impact of residential electricity storage at local and grid level for Portugal. *Applied Energy, 128*, 254–264. https://doi.org/10.1016/j.apenergy.2014.04.054
56. Resch, M., Bühler, J., Schachler, B., Kunert, R., Meier, A., & Sumper, A. (2019, January). Technical and economic comparison of grid supportive vanadium redox flow batteries for primary control reserve and community electricity storage in Germany. *International Journal of Energy Research, 43*(1), 337–357. https://doi.org/10.1002/er.4269
57. Le Cadre, H., Jacquot, P., Wan, C., & Alasseur, C. (2020, April). Peer-to-peer electricity market analysis: From variational to Generalized Nash Equilibrium. *European Journal of Operational Research, 282*(2), 753–771. https://doi.org/10.1016/j.ejor.2019.09.035
58. Gunarathna, C. L., Yang, R. J., Jayasuriya, S., & Wang, K. (2022, July). Reviewing global peer-to-peer distributed renewable energy trading projects. *Energy Research and Social Science, 89*, 102655. https://doi.org/10.1016/j.erss.2022.102655
59. Biggar, D., & Hesamzadeh, M. R. (2022). Energy communities: Challenges for regulators and policymakers. In *Energy communities* (pp. 131–149). Elsevier.
60. Hall, S., Anable, J., Hardy, J., Workman, M., Mazur, C., & Matthews, Y. (2021, April). Matching consumer segments to innovative utility business models. *Nature Energy, 6*(4), 349–361. https://doi.org/10.1038/s41560-021-00781-1
61. Wilkinson, S., Hojckova, K., Eon, C., Morrison, G. M., & Sandén, B. (2020, August). Is peer-to-peer electricity trading empowering users? Evidence on motivations and roles in a prosumer business model trial in Australia. *Energy Research and Social Science, 66*, 101500. https://doi.org/10.1016/j.erss.2020.101500
62. Herbes, C., Brummer, V., Rognli, J., Blazejewski, S., & Gericke, N. (2017, October). Responding to policy change: New business models for renewable energy cooperatives – Barriers perceived by cooperatives' members. *Energy Policy, 109*, 82–95. https://doi.org/10.1016/j.enpol.2017.06.051
63. Tushar, W., et al. (2021, January). Peer-to-peer energy systems for connected communities: A review of recent advances and emerging challenges. *Applied Energy, 282*, 116131. https://doi.org/10.1016/j.apenergy.2020.116131
64. Peer Energy Cloud. (2020). http://software-cluster.org/projects/peer-energy-cloud/. Accessed 01 Sept 2022.
65. Piclo Building a smarter energy future. (2020). https://piclo.energy/about#whitepaper. Accessed 01 Sept 2022.
66. Powerpeer – Power to the people. (2020). https://www.powerpeers.nl/login. Accessed 01 Sept 2022.
67. Mengelkamp, E., Gärttner, J., Rock, K., Kessler, S., Orsini, L., & Weinhardt, C. (2018, January). Designing microgrid energy markets. *Applied Energy, 210*, 870–880. https://doi.org/10.1016/j.apenergy.2017.06.054
68. Australian Renewable Energy Agency AGL virtual trial of peer-to-peer energy trading. (2020). https://arena.gov.au/projects/agl-virtual-trial-peer-to-peer-trading/. Accessed 01 Sept 2022.
69. Power Ledger RENeW nexus Australian government Australia. (2020). https://www.powerledger.io/project/renew-nexus/Google Scholar. Accessed 01 Sept 2022.
70. Power Ledger KEPCO Japan – Peer-to-Peer solar power and REC trading. (2020). https://www.powerledger.io/project/kepco/. Accessed 01 Sept 2022.

71. Power Ledger Sharing energy & erex, Japan peer-to-peer solar power trading. (2020). Accessed 01 Sept 2022.
72. Power Ledger Uttar Pradesh Government, India – Peer-to-peer solar power trading. (2020). https://www.powerledger.io/project/up-government/. Accessed 01 Sept 2022
73. Axsen, J., Orlebar, C., & Skippon, S. (2013, November). Social influence and consumer preference formation for pro-environmental technology: The case of a U.K. workplace electric-vehicle study. *Ecological Economics, 95*, 96–107. https://doi.org/10.1016/j.ecolecon.2013.08.009
74. Singh, A., Strating, A. T., Romero Herrera, N. A., Mahato, D., Keyson, D. V., & van Dijk, H. W. (2018, December). Exploring peer-to-peer returns in off-grid renewable energy systems in rural India: An anthropological perspective on local energy sharing and trading. *Energy Research & Social Science, 46*, 194–213. https://doi.org/10.1016/j.erss.2018.07.021

# Chapter 2
# Introduction and Use Cases of P2P Trading

**Bahman Taheri, Farkhondeh Jabari, Asghar Akbari Foroud, and Reza Keypour**

## 2.1 Introduction of P2P Energy Exchanges and Transactive Energy (TE)

After some time of relative stability in the industry and according to the recent developments in the energy sector, the power system was able to introduce a new model of energy distribution called Peer-to-Peer (P2P) trading system for communities, and the effectiveness of this trading system in benefiting all actors have been proven in multiple studies. Recently, P2P trading has been suggested as a novel mechanism to strengthen the straightforward sharing of electricity between players in the multi-level market system with responsibility and predetermined privacy. However, compared to other P2P asset markets, this system has created a lot of problems due to the obstacles to the cooperation of non-professional actors and regulated actors such as distributed system operators and transmission system operators. There is no doubt that P2P energy trading has been able to revolutionize the power sector with next-generation energy management with the help of existing systems.

Also, the research work and implementation of P2P business projects in the past years have led to economic, social, and technical commonalities and methods for progress in the following aspects [1].

B. Taheri · A. Akbari Foroud (✉) · R. Keypour
Electrical and Computer Engineering Faculty, Semnan University, Semnan, Iran
e-mail: b.taheri@semnan.ac.ir; aakbari@semnan.ac.ir; rkeypour@semnan.ac.ir

F. Jabari
Power Systems Operation and Planning Research Department, Niroo Research Institute (NRI), Shahrak Ghods, Tehran, Iran
e-mail: fjabbari@nri.ac.ir

 
V. Vahidinasab, B. Mohammadi-Ivatloo (eds.), *Demand-Side Peer-to-Peer Energy Trading*, Green Energy and Technology, https://doi.org/10.1007/978-3-031-35233-1_2

- Physical integration of energy systems
- Flexible services at different scales
- Coordination between multi-level markets

Unfortunately, in the academic and industrial fields, limited attention is paid to the research done on P2P business, and the existing approaches regarding issues and future research have only been examined and studied in articles, which include the following.

- Existing challenges and legal aspects
- Market models and design
- Key technologies
- Trading platforms, projects, and startups
- Actors
- Business trust, security, and privacy

In order to successfully create a P2P energy community and have security and privacy, actors and participants should use interdisciplinary integration of different fields. The P2P trading system is presented as a platform that can create a convenient process for independent decision-making of energy buyers. In this process, a customer can decide completely independently about the energy parameters he needs, for example, how much energy to share or how to set his price, and even when and when to sell. In energy systems, a four-stage structure for P2P energy trading has been presented and proposed. This structure is presented in Fig. 2.1 and the constituent elements and the main technologies involved are defined based on the roles they have and play in this process. There exist three main dimensions to this structure [2]. The main steps and cases related to energy trading systems can be divided into four steps, which are shown in Fig. 2.1.

- The power grid ladder includes all the physical and main components of the power system, including feeders, power transformers, smart meters, electric loads, DERs, etc., which are the main components and constituents of the physical distribution of electricity in which P2P trading is implemented.
- The ICT phase also includes communication systems, protocols, applications, and information flow.
- The control stage mainly includes the control functions of the power distribution system. In order to maintain security and increase the reliability of the power supply and power flow control, various methods have been determined to control these items, including voltage control, frequency control, and active power control.
- In these markets, communication devices include things like sensors, wired/wireless communication links, routers, switches, servers, and various types of computers.

Recently, a P2P economy known as the sharing economy has been proposed for how P2P trading systems are offered. Each peer (consumer, producer, buyer, or even supplier) in P2P energy trading is allowed to choose a peer for energy trading

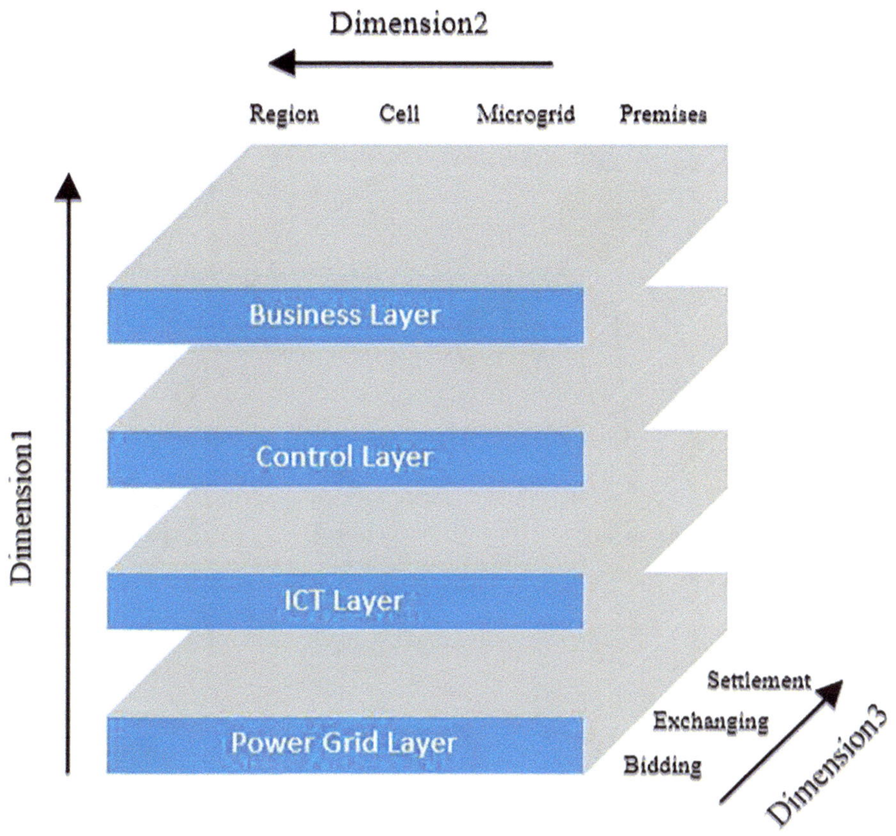


**Fig. 2.1** A 4-step system structure of P2P energy trading

(actors) according to its objective, for example, maximum profit, reliable energy supply, minimum cost, minimum pollution, etc. In order to avoid possible future issues, it should be done by learning from past experiences and also by considering both general and scientific aspects of a complex multidisciplinary paradigm including TE. Except for these cases, the methods and tools used in integrated systems should be used. Current can be necessary to check [3, 4] (Figs. 2.2 and 2.3).

From DSO's point of view, the high penetration of DERs, especially renewable sources, is evident. Their intermittent nature leads to network management issues. But considering that the consumer can use the generated electricity to reduce the electricity bill, DERs are interesting programs from the customer's point of view. In this context, a control method to integrate the high influence of undercurrent DERs in the system called TE is presented to increase the safety and make the network more efficient. However, it is possible to continuously meet the manufacturer and the customer in P2P. While TE control method is defined in a P2P market equipped with DRER for sharing and trading of energy between all consumers, which is connected

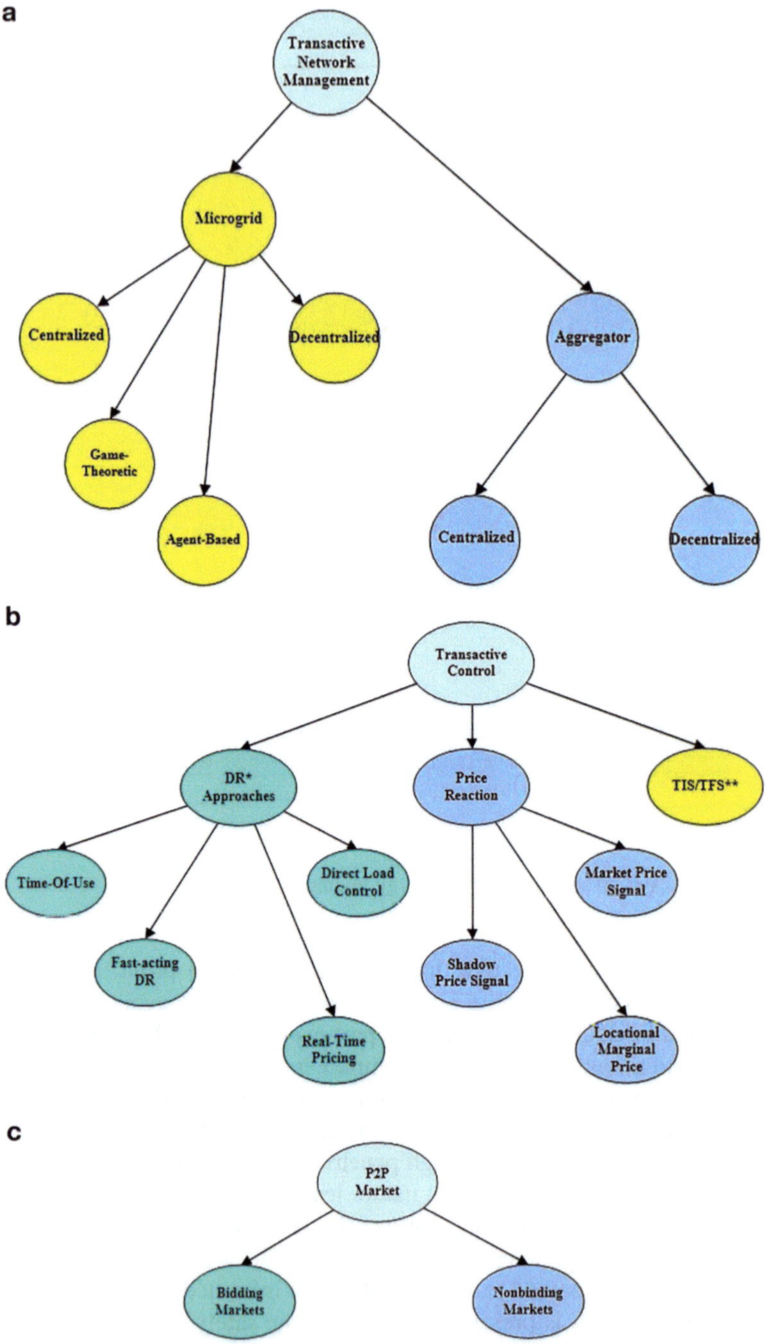


**Fig. 2.2** Concepts and classifications related to TE (**a**) transactive network management, (**b**) transactive control, and (**c**) P2P markets

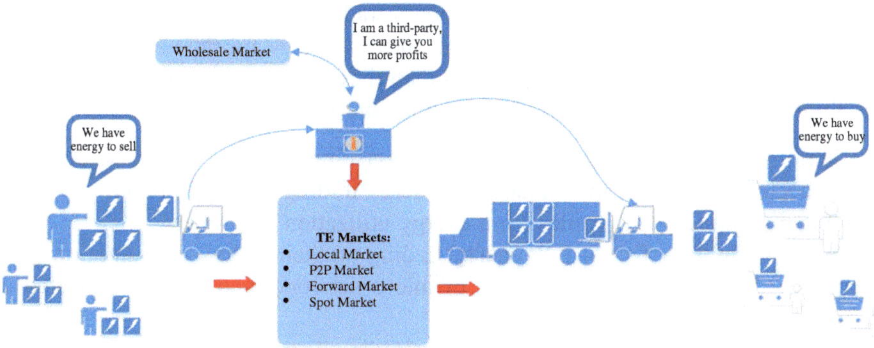


**Fig. 2.3** TE system implementation steps [1]

by buying/selling energy from each connected node. This network turns into active customers (sellers) in the market. Therefore, one of the most promising paradigms for the market can be considered the TE system and P2P trade, which are completely connected.

Note that all processes must be implemented at the network distribution level. In other words, direct energy sharing between consumers and/or sellers in P2P energy trading is only possible in local power grids. Currently, the concepts of P2P markets have been explored through several research and industry projects, focusing on this business at the distribution level to integrate all small and medium-sized DERs. To have a fully decentralized market in a P2P market, each buyer has their own local controller, and local decisions are made based on users and market information. Consider each seller in P2P markets a Transactive Node (TN), and all TNs participate in this business. Obtaining optimal and cost-effective performance is necessary considering several constraints for the market offer and business partner selection. In P2P markets, the buyers are the same consumers who exchange surplus production with those who demand energy. This energy business is carried out through several long-term or temporary contracts between all layers of the network. Two types of contracts are proposed by [5]: (1) between energy sellers (e.g., one seller exchanges generated electricity with another seller); (2) between suppliers and consumers of energy (e.g., one actor is only responsible for production and the second actor is only a consumer). Sharing information and exchanging energy together in the P2P market is similar to the concept of the Internet. Equivalent nodes called "clients" or "servers" exist throughout the Internet. This network is able to exchange information and data between Internet networks using nodes because nodes are both clients and servers at the same time. This fact is also valid in P2P. In the P2P market, energy players can exchange information simultaneously.

In order to implement P2P energy trading systems, another layer of software is necessary in addition to the hardware and infrastructure requirements. It also provides the network operator with control and monitoring of energy trade and reduction of data transmission between all system actors by a software platform.

ELECBAY platforms [4, 5] allow P2P energy trading in a microgrid. This platform provides a list of products by sellers (e.g., surplus energy for the next 30 minutes), which can also be viewed by buyers, and then they register their order according to the time and place. This platform is important and used for:

- Demonstrating how consumers and/or consumers of energy in a network-connected LV microgrid conduct P2P energy trading with each other.
- Obtaining novel load profiles from energy utilization and sellers.
- Determining the effect of P2P energy trading on their energy consumption.
- Empowering the analysis and rein of micro networks and power connected to the grid.
- The bidding strategy through P2P energy trading determines the supply of energy consumers and consumption values before the gate is closed. The production profile of energy consumers and the consumption profile of consumers and energy consumers are based on forecasted information.

## 2.2 Market Players, Structure, and Optimization Models

P2P energy community players are usually from different levels of the market, which are shown in Fig. 2.4. For example, market sellers can change their role to buy or sell electricity by choosing their business goals more flexibly based on the status and price of the transaction. Various P2P electricity business models and mechanisms have been proposed by researchers, which can guide all the trading markets presented in Fig. 2.4 and can form a P2P energy community. Various roles and responsibilities are chosen by actors in P2P business, and they are involved, and in order to form a strong P2P market, the following issues should be addressed:

- Extensive communication to exchange flexibility between sales and customer departments
- Interoperability and integration of heterogeneous systems of the Internet of Things on large scales and with the participation of actors
- Establishing trust and interaction between actors without the involvement of a third party by P2P
- Consumer/product-oriented markets with customized applications and automated business processing

The traditional electricity network to support the flow of electricity is usually built in a one-way manner, and currently, due to the expansion of new renewable energy resource technology, electricity markets have become more modern and up-to-date markets. And this is because all customers are motivated to invest in renewable energy systems to be able to meet part of their demand, and in new local systems, consumers who are equipped with DER will be able to meet part of their demand. Consumption of self-produced electricity and also consumption by energy storage devices, store or even export that energy. Therefore, for the reasons mentioned, they

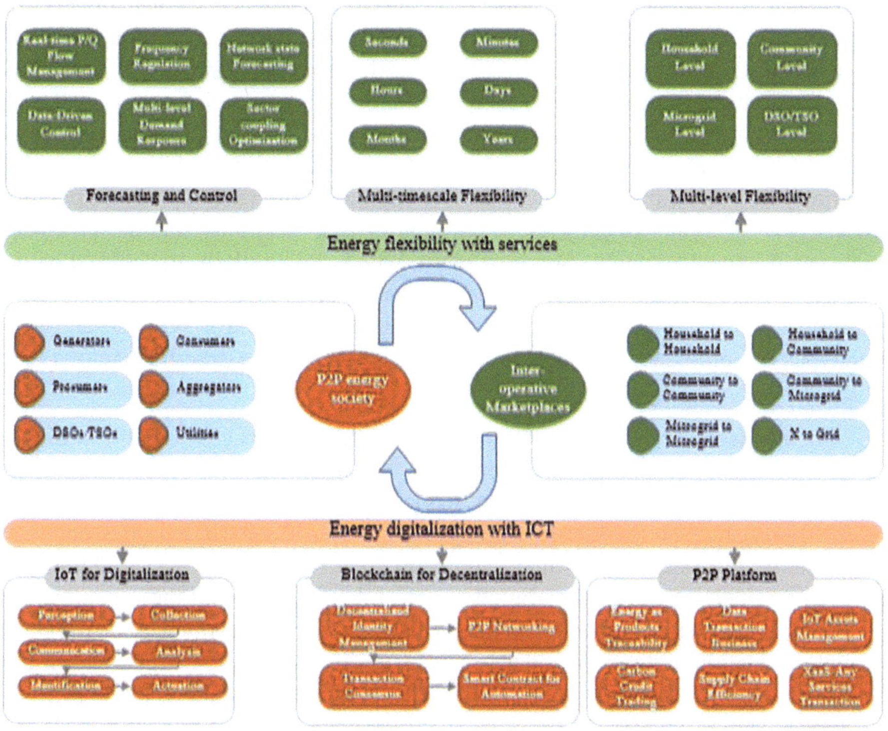


**Fig. 2.4** P2P energy community players

can be called consumers and producers at the same time, and the basic concept is that energy sellers will be able to sell their produced electricity directly to others, even at a higher price, and the buyer can also buy electricity directly. Buy what you need from others and at a much lower price.

In the last few years, researchers have shown great interest in the study of trade P2P because the trade market P2P can be used as a way to combine the use of new energies while also greatly reducing pollution. The weather must be effective and that is why many projects are implemented around the world.

The overall goal of P2P trade and market is to create social welfare among energy players and especially in this type of market where multiple players can compete and cooperate freely. In previous studies, researchers have also proven the effectiveness and benefits of P2P for the participants and actors of this business. However, the creation and development of P2P markets is a very difficult and complex task, and so far, many studies have been done in this field and have been able to satisfy their players. In general, P2P business is displayed in the following categories.

- Intensive trade market
- Decentralized markets
- Distributed markets

In distributed markets, import and export orders are not usually given directly and they affect the market only by sending different prices. In fact, distributed markets can be presented as a solution between centralized and decentralized markets. P2P markets can be considered as a fundamental solution to satisfy network actors (consumers and sellers). Among consumers equipped with DER, the P2P market is considered as electricity trading and actors can easily and freely buy or sell the electricity they need from wherever they want and become active customers in the market [1]. A number of projects in the field of P2P electricity trading have been analyzed in recent research. Some of them, which act similarly to the role of the supplier in the electricity sector, are focused on business models and energy market platforms, and others are focused on micro-grids, local control systems, and ICT. The types of optimization models in peer-to-peer trading are described below.

- *Piclo*

  It was used for the first time by English researchers. "Open Utility," which is an innovative technology company, and "Good Energy," a renewable energy supply company, have done joint work in which consumers could buy the electricity they need directly from the local producer. All required data and consumer information for supply and supply are modeled for every 30 minutes. In this model, the generators directly control the purchase and the consumer can also prioritize which generator to buy. This planning first reconciles consumption and production and provides data analysis to the customer. Also, the renewable energy supplier company balances the market according to the contract [6].
- *Vandebron*

  This platform is used in the Netherlands, and in this type of application, the consumer directly buys and supplies the electricity demand from independent generators. For example, farmers who buy and supply the electricity they need from the wind turbines in the fields. In fact, this model also acts as a supplier and balances the market by linking consumers and producers [6, 7].
- *Peer Electricity Cloud*

  It was launched as a research project in Germany and used cloud-based technologies to deal with the overproduction of a local e-commerce platform. Also, innovative methods of recapture and prediction for device-specific power consumption have been used to create a virtual market for electricity trading and construct value-added services in the microgrid.
- *Smart Watts*

  It was a scientific project in Berlin, Germany, which was successfully implemented and presented and tested new and optimal approaches to electricity supply through new and updated information and communication technologies. It has also used the potential of information and communication technology to provide more energy with high security [7].
- *Yahola and Mosaic*

  Both Yahola and Mosaic models were presented in America. These programs allow consumers of apartments and those who do not use solar electricity to meet a part of their energy needs through the existing solar system and reduce the water bill and reduce their electricity so that they can save their costs even if they change their place and move. These programs are also similar to Piclo and Vandebron but use solar energy instead of renewable energy.

- *Sonnen Community*

  The Sonnen forum was developed by Sonnen batteries in Germany by battery manufacturers; it is a community of battery owners who can share self-generated electricity with others. As a result, with such a system, there is no need for production. By using this type of battery and on the other side of a PV system, you can fully supply the energy you need. This excess electricity is not supplied into the regular power system but into a virtual energy pool that serves other members when they cannot generate enough power due to bad weather.

  By using a program and software, all consumers and members are connected to each other and monitored, and supply and demand are balanced. The idea is very similar to Piclo's and Vandebron's, but Sonen Community can easily reveal the importance of storage systems.

- *Lichtblick Swarm electricity*

  Swarm electricity is a collection of services provided by energy supplier Lichtblick. Swarm Conductor, a part of Swarm electricity services, is a unique type of scheduling in the electricity market. In this planning, all up-to-date processes that are liked by electricity customers, including residential and commercial customers, have been combined. In this production planning, improved batteries and Swarm electricity have enabled the interaction of renewable and distributed electricity sources [8].

- *Transactive Grid*

  Transactive Grid is a combined market of software and hardware, whose members can enjoy significant safety and automatically purchase energy from each other by using smart contracts and blockchain. The prototype uses the Ethereum blockchain in Brooklyn, where consumers can determine for themselves where and from which vendor they get the electricity they need. Home electricity producers are allowed to sell their surplus to neighbors and others, and communities can keep energy sources local, reducing waste and increasing micro- and macro-grid efficiency [1].

- *Electron*

  Electron is a revolutionary new platform for gas and electricity metering and billing systems, which is still under development. It will open the way for exciting and innovative consumer energy services. It is a completely secure, transparent, decentralized platform that runs on a blockchain and provides a provably honest metering, billing and switching service using Smart Contracts and the power of Distributed Consensus. The platform will be open source and operate for the benefit of all users. It will not be owned or controlled by suppliers or brokers [9] (Table 2.1).

Table 2.1 Research plans and their comparison with each other [7]

| Project name | Country | Start year | Objectives | Network size | P2P layers | Outcomes | Shortcoming |
|---|---|---|---|---|---|---|---|
| Piclo | UK | 2014 | P2P ETP from suppliers perspective | National | Business | A P2P ETP | No discussion on local markets |
| Vandebron | Netherland | 2014 | P2P ETP from suppliers perspective | National | Business | A P2P ETP | No discussion on local markets |
| Peer energy cloud | Germany | 2014 | Cloud-based P2P ETP, smart home | Micronetworks | Energy network ICT | Cloud-based platform for smart homes | No discussion on control system |
| Smart watts | Germany | 2011 | Optimizing energy supply via ICT | Regional | Energy network ICT | A smart meter gateway as interface to Internet | No discussion on control system |
| Yahola and Mosaic | US | 2015 | Solar sharing network for lower energy bills | Regional | Business | Terminated due to funding issues | No discussion on local markets |
| Sonnen Community | Germany | 2015 | P2P energy trading with storage system | National | Energy net-work business | A P2P ETP | No discussion on local markets |
| Lichtblick swarm energy | Germany | 2010 | IT platform for energy mar-kets and customers | National | Energy net-work ICT | Plenty of services provided by the energy supplier | No discussion on local markets |
| Transactive grid | US | 2015 | P2P energy trading within micronetworks | Grid-connected micronetworks | Energy net-work control, ICT, business | Automatic energy trading platform within microgrid | No discussion |
| Electron | UK | 2016 | Energy metering and billing platform using blockchain | Unknown | Energy Net-work, ICT, business | Not started yet | Not started yet |

## 2.3 Advantages, Disadvantages, and Challenges of P2P

The paradigm of electricity markets in today's systems has changed from a centralized structure to a decentralized structure due to the daily increase in renewable energy production. One of the new architectures of the decentralized electric power market is in the power and distribution systems of the market microgrid. This type of business allows neighboring customers who can generate electricity to exchange and sell their generated energy directly to each other. The full history of the energy system (P2P) is detailed in Ref. [10]. The blockchain system can become a precise tool to support the payment system and secure the private information of the players of this business. The idea of using the blockchain system in the market (P2P) has been upgraded from a hypothesis and plan to a practical system in the past years. This pricing plan system plays a critical role in the execution of transactions. There are many challenges in the existing systems, and to solve them, a new system should be presented. The P2P trading as a new system allows a direct interaction between producers and consumers in the market without considering third parties [4] P2P is the ability to trade electricity with each other (consumers or consumers), earn revenue for surplus power, use a low-cost settlement system to reduce electricity bills, and improve the return on investment in distributed generation. The P2P trading opens up the possibility of switching energy suppliers on a minute-by-minute basis and buying and selling electricity (vendors) based on your preferences. For example, a P2P system might use blockchain technologies to track the amount of electricity traded and have a transparent automatic settlement system [8]. However, the concepts of P2P energy trading in micronetworks are still at a high level. Until now, there is no consensus among researchers about what type of market or what business model or market design can help the development of local markets [11].

In this regard, digitalization of new systems makes P2P trade possible and facilitates the creation of local markets, which can ultimately improve the local electricity markets. This issue has led to various questions in the design of markets, including the questions of which local electricity market plans and P2P mechanisms provide a suitable framework for efficient exploitation of the digitalization of electricity distribution networks [12]. In P2P energy trading, small-scale markets are usually used in offices, factories, and homes.

### *2.3.1 Peer-to-Peer Advantages*

A P2P approach to energy trading and sharing would promote energy availability in the community and increase efficiency, flexibility and effectiveness of local resources. Considering that the P2P market has many advantages because every consumer who can sell his surplus production gets financial benefits, and instead of buying directly from the network, consumers buy from other consumers at a lower price. Also, by implementing markets (TEM) in smart networks, it enables flexible programs to be implemented in which producers and consumers can freely plan production and consumption of DER based on the consumption required by institutions and departments. Various actors, including retailers, change to provide network services during the period [13]. All system operators must be able to cope with the ever-increasing complexity of the network as well as coordinate energy consumption and production, which can only be done with the help of TEM.

### *2.3.2 Disadvantages and Challenges P2P*

Among the disadvantages of this type of business, we can point out things such as the novelty of the subject to the market and customers. Also, academic researchers have not focused significantly on this business to be able to introduce this business to actors.

## 2.4 Applications of Blockchain Technology in Transactive Energy Systems

Blockchain was first used by several countries to promote local electricity production and consumption and electricity distribution and decentralization. A significant improvement in the functioning of electrical systems was created with the creation of decentralized and secure communication systems. A reliable, open, and decentralized record for all data and transactions related to electricity production and consumption can be created due to the nature of blockchain technology. Due to the decentralized and digital nature of the technology, blockchain can be used to make the flow of data between participants and various components in the energy system safer and more transparent while simplifying operations [13, 14].

The installation of DG technologies, especially PV, has turned ordinary consumers into active players in the local supply of electricity. These developments, along with the digitization of distribution networks (smart grid), have provided the arena for reaching a new model called P2P electricity trading. However, the design of features and roles for how to buy or sell electricity locally for microgrids or small communities is in its early stages. Market design studies mainly focus on established and mature electricity markets and have not focused so much on encouraging local trading. This is partly because the concepts of local markets have distinctive features: the diversity and characteristics of distributed generation, specific rules related to local electricity prices, and the role of digital tools in facilitating P2P transactions.

The work protocols of the distributed computer network of blockchain technology are said to be able to securely manage and maintain the data entered and processed by users without the need for a centralized authority. Although this definition is a general definition of blockchain, it may encompass the most advanced applications that support the disruptive transformation of many sectors, including finance, transportation, supply chain, and most recently, the energy sector [12].

## 2.4.1 State-of-the-Arts of Blockchain Applications

The intermittent nature of renewable energy sources as well as their unpredictability in power systems may lead to unanticipated production peaks that will be different from demand. Therefore, in order to properly manage this inconsistency, proper communication between consumers and active users is necessary. In order to achieve this goal, the approach of traditional power systems should no longer be used because the actors on both sides, both the customer and the active user, want to participate in energy exchanges, and a decentralized system and technology based on TES and P2P energy should be adopted.

The secure proposal in this field will be Distributed ledger technologies (DLT), which work as the most promising solution for smart contracts between the players of this market and are based on the blockchain concept. An overview of how decentralized TESs are deployed is presented in [15]. Also, in this article, how to use the proposed plan in a virtual power plant and taking into account the collectors and consumers of smart buildings equipped with a transaction controller to manage the battery system has been fully discussed. Advanced blockchain technology can provide a shared decentralized ledger with high security for individuals and companies to exchange information and is fully described in [16]. The basic features of the combined blockchain are also fully presented and described in [17] in terms of calculation and theory and technical and economic principles.

## 2.4.2 Standards and Protocols of Blockchain

Due to the fact that blockchain programs are growing very fast, there is an urgent need to standardize the technology and update it to optimize their interoperability and use. The development of standards is subject to oversimplification of the field. Especially the numerous suppliers that promote the market blockchain. A lot of effort and research should be done to create two-way communication between market players and the goals of using consensus algorithms such as star consensus and chart should be considered for standardization.

### 2.4.2.1 Standardization ITU

The United Nations specialized agency ITU makes a major contribution to international information and communication technologies (ICT). And the main role in analyzing the technical, tariff, and operational areas and providing recommendations with a standard perspective at the global level is held by ITU-T. Work cases in this system can be divided into three categories, which are fully presented in [18].

#### 2.4.2.2 ISO

This item was established and presented in 2016 and it has six working groups (WG) including six different WGs, and a research group (SG) named SG 7, which is fully focused on the security of applications and privacy of actors, respectively, smart contracts, blockchain, and techniques. Information technology security, governance, use cases, and interoperability of DLT [19].

#### 2.4.2.3 IEEE Standards Association

Related study system DLT that are developing IEEE system.

#### 2.4.2.4 W3C

The draft contract and document published by social activists and verified by the blockchain community is presented in [19].

#### 2.4.2.5 United Nations Center for Trade Facilitation and Electronic Business (UN/CEFACT)

UN/CEFACT is an organization under the auspices of the United Nations for Europe (UNECE), which has recently provided general regulatory frameworks and international standards for emerging technologies, including blockchain. UN/CEFACT has published white papers on two criteria, namely: the introduction of blockchain technology especially for trade policy makers, and research gaps in the technical aspects of blockchain in relation to UN/CEFACT delivery. Also, in order to be used against the existing challenges and facing government services that are in the research stage and extensive research should be done to achieve their goals [19].

#### 2.4.2.6 Community Standards

Society and industry organizations have begun to more actively promote DLT standards. Official standards development organizations that have published different implementations of the specification together with an open-source implementation code repository. The Linux Foundation's blockchain programs are an example of a community-driven organization that supports a wide range of industries and is hosted by the Linux Foundation's hyperledger, which focuses on about 200 companies. One of the works of hyperledger is considered as a real standard for blockchain development and implementation. Offerings (EIP) are continuously provided by the Ethereum community through a process called Ethereum Proof of Stake. The (ERC) standards include protocol specifications, programming interfaces, interface/client programming, and application-level standards. The most important ERC-20 standard is used in smart contracts on the Ethereum blockchain, which is available for token implementation. In order to develop the industry, widely used standards such as DLT are used. In August 2017, the Blockchain Transport Association (BiTA) was established to meet such needs, which has concentrated standardization efforts with regard to the transportation industry [19, 20].

#### 2.4.2.7 Other Standards

An authorized distributed ledger has been launched by the Telecommunication Standards Institute and ISG Industrial Group, which includes three work items:

- The future of standards and technologies
- The ability to implement
- Compliant with data processing requirements for connected machines and their field of application.

### 2.4.3 *Building Trust and Matching the Blockchain and its Standard*

When it is necessary to use RFID (radio frequency identification) to enable tracking of items that are produced in high volume, no one can use the same registry tag that is assigned to another department. Other and miscellaneous tags are commonly used to replicate information, and distributed ledgers can replace physical tags, but there is still an overlap between the two.

### 2.4.4 *Blockchain Standards and Government*

In general, it is based on four important principles that the government can use to invest in blockchain, which is a distributed ledger that is encrypted using different keywords and is immutable and democratic. In fact, based on these principles, blockchain can provide low-cost solutions to solve problems. Also, if necessary, traditional databases will help meet this need. A very unique solution will be provided through the blockchain system. Consensus requirements are important for government use of blockchain-based systems, which are unlikely to be acceptable to the US government, while none of the existing government entities are blockchain independent to any degree in relation to their peers. In any contract that is usually concluded, one of the basic and special pillars is trust between the parties. One of the basic programs offered in America by blockchain technology is the temporary work visa, which has been able to employ several thousand workers in various fields in this country. In order to implement this program, employers must send their application and application for temporary work certification to the Department of Labor. And after doing these things, the employer must send the application to the US Citizenship and Immigration Services. After dealing with the employer, the worker has to submit his visa application to the USCIS advisor after USCIS approval. One blockchain application is an experiment that explores many ways to make it more efficient for employers. To track and manage physical assets, a pilot will be launched by the Treasury Department's Office of Financial Services Innovation and Transformation (FIT). By using the pilot control and regulation plan, the inventory of physical assets of the agency can be examined. In 2018, in order to create applications based on the blockchain, centers were established, which include the disease control center, the monitoring center, and laboratory services. In order to make better and more efficient use of blockchain technology in managing data during a public health crisis, establishing better health monitoring will be the most important task of the point-of-care (PoC). Currently, the public service department is working on the basis of machine learning and artificial intelligence [21].

## 2.5 Managing Interactions Between the Aggregator and the Buyer by Deploying a Smart Contract

New opportunities are provided by using decentralized energy system management for safer, more flexible, and more efficient energy distribution. It is true that using these opportunities is promising, but these systems are currently in the early stages. In P2P trading, buyers are able to cooperate to share information about energy demand and supply. A P2P marketplace can run without the need for a central controller and any failure in its subsystems/programs. This allows easy replacement of any of its subsystems without changing the control modules. In order to check the performance of P2P, it is necessary to start at the microgrid level by exiting the features of the subsystems of the microgrid programs in order to check the correct operation of the trading system and energy sharing. If the architecture made for P2P communication matches the required characteristics of microgrid programs, the integration of P2P and its architecture with business and energy sharing will be easily integrated. The energy data generated by each program is transmitted to an endpoint via two-way communication when needed so that errors can be detected in time. In a typical microgrid, three levels of two-way communication are introduced. The first is the energy generation and utilization infrastructure (DERs, loads), the second is to control the microgrid control center, which is all the components of the program, and the third is the microgrid communicates with other neighboring microgrids [22].

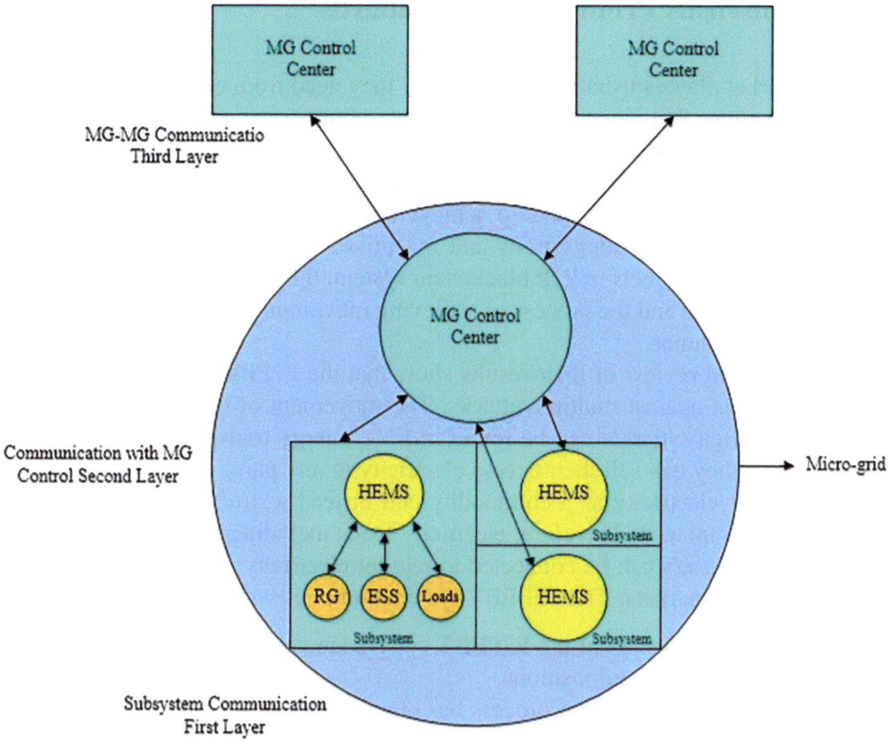


**Fig. 2.5** Hierarchy and communication layers in the network

According to Fig. 2.5, it is necessary to evaluate the characteristics of microgrid applications in order to evaluate the compatibility of P2P communication architectures to control all applications and ultimately for customers who share and trade energy in P2P mode. P2P trading increases the efficiency, flexibility, and effectiveness of local resources, and for trade and energy sharing, it can also improve the availability of energy in the community. According to the mentioned cases, some research projects, such as Piclo in London, Sonnen in Germany, and Vandebron in the Netherlands, have been started [10]. Network planners can use P2P schemes to organize and maximize network subsystems and maximize network efficiency. The classification of the architectures of these markets is basically based on how information is routed in the network system and how other actors are placed. In the centralized information platform, a requesting peer asks a specific central entity to store a list of all resources for information (or IP addresses) of other peers that own a specific resource. The requesting peer receives the information of the source peer and direct communication is established between both peers without passing through the central entity [16].

## 2.6 Consensus Protocols and Standards

Today, market players usually get the energy they need from centralized companies or organizations that do not have sufficient security against cyber-attacks. The authors have reported an intelligent power trading system based on a peer-to-peer consensus protocol (P2PEBT) in [21]. Considering the increase of electric vehicles (EVs) integration in the network to achieve demand response, P2P system can be provided by providing suggestions and incentives for local electricity balancing. Using the smart contracts in P2P blockchain system, this plan is implemented on the Ethereum platform and the process of achieving maximum benefits is completed in a decentralized manner.

Research and review of their results show that the P2PEBT system can provide good protection against multiple attacks. The movement of market players toward the P2P electricity market can be found in EVs, energy transport vehicles, in fact, this is where they can sell their excess electricity to any party they want. They may even sell their electricity as a commodity and instead a structure New and hierarchical will be replaced. Therefore, electricity loads including two main commercial and residential users can be connected to regular electricity generators in retail and even wholesale markets. The P2PEBT system complies with the following rules:

- The basic components of the P2PEBT system are considered as the reference of the current market compositions.
- Efficiency: The transaction process has eliminated third-party intervention compared to the traditional retail process, thus reducing the amount of time spent and improving efficiency in a transparent transaction.
- Flexible: In this model that uses an open trading platform, more customers and buyers are allowed to enter the exchange market more easily.
- Economic: Considering that the new power system can create a direct connection between market users and also ensure the overall stability of the power grid, on the other hand, the new technology can bring economic efficiency to both the grid operators and the Maximize single users.

## 2.7 Impact of Cyber-Attacks on P2P Energy Exchanges and Transactive Energy

Recently, TEM has been used to balance demand and supply across the grid due to the increasing integration of distributed energy production in the power grid. TEM can create flexibility by providing P2P and reducing user demand in the network, it can also increase system efficiency and reduce pressure on the energy network. Considering that the actors in this business are equipped with smart devices and participate independently in the energy market and exchange a large amount of information through the communication channel, the P2P trading system will be vulnerable to cyberattacks. Therefore, in order to have a system resistant to these attacks, the cases must be fully investigated and analyzed in order to design and create a resistant system based on TEM. The purpose of this section is to identify cyber-attacks related to a TEM-based power system. In order to achieve this goal, P2P attacks must be accurately classified. In traditional systems and markets, although TEM has provided advantages for new and modern systems with the integration of DERs, on the other hand, it has created an opportunity to attack cyber attackers and security threats, which never happened in old networks [1].

In this part, it is tried to discuss a number of FDIAs that are used in P2P energy trading and daily flexibility schemes of the electricity market in microgrids as a TEM proposed by researchers for energy management in microgrids. They usually provide operators and users with false data, such as offers and price signals, demand/supply, to increase their personal interests. According to the explanations provided, TEM includes implementations such as intelligent devices, users, communication channels, and market operators, etc. Communication channels, smart devices, and unwanted users, among the market components can act as the main part for implementing cyber-attack programs due to the fragile security mechanism and direct access to the system. Therefore, the attacks can be classified into three main parts based on the implementation of the TEM market:

- Conducting a cyberattack through communication channels
- Carrying out attacks through existing devices
- Carrying out attacks through malicious network users

It should also be mentioned that malicious users and enemies usually carry out cyberattacks on communication devices and channels while attacking and malicious users act in the third category.

## 2.7.1 *Cyber-Attacks with Communication Channels*

The information required during energy trading and information transfer flexibility is usually done through a communication channel and a component in TEM. In this trading market, the supply and demand of energy as well as the information of TAs are transferred to the market operator through a communication link [6] (Table 2.2).

**Table 2.2** Attacks and their impact on the market (TEM)

| Attacks | Benefits | Reducing the well being of users | Market disturbance | Bring catastrophic events |
|---|---|---|---|---|
| Attack through communication channel | + | + | + | + |
| Attack through devices | + | + | + | + |
| Attack through adverse users | + | + | + | + |

### *2.7.2 Cyberattacks on Existing Devices*

Some consumers, such as TA in TEM, can create system security in a special way in the network. Usually, the enemy tries to find items such as household appliances, heating and cooling systems, as well as air conditioning systems, which have lower security than others, and subject them to cyber-attacks. Considering that implementing of TEM requires a wide infrastructure and highly advanced measurement in network systems, where a smart meter is installed at the end of each TA to measure electricity consumption in real time. And on the other hand, due to the fact that the smart meters in the network and the society are connected to each other through wired or wireless networks, cyberattacks are usually carried out on smart meters [9].

## 2.8 Challenges and Future Trends in P2P Energy Markets and TE Systems

New models have been introduced by P2P energy markets to trade electricity locally. In the past years, many advances have been reported by the study of academic researchers in the logic of P2P trading markets. This type of business and market has been a very good alternative for actors and energy buyers in recent years. Therefore, P2P is summarized as a design and transformation of electricity business in energy bidding. Also, it is no longer the human who is the decision-maker in new systems and new energy trading markets, and today's advanced technologies have been able to speed up these markets and transactions. Because in these markets, due to the uncertainty of RES output load and customer demand, it becomes a supply-and-demand relationship, a software-based system will be needed to enable effective and efficient negotiations at the moment. Representatives in this system must be well-behaved, well-mannered, and intelligent while negotiating with others. A challenge in this field will be to have an effective, acceptable, and efficient design so that others can use this design to guide their suggestions, which researchers have made a lot of efforts in this field. P2P energy trading allows consumers to trade and sell their excess energy production with others and increase their income. Also, this market has shown more flexibility to its players and has given an opportunity to use renewable and clean energy, which can be effective in minimizing air pollution in the region. Also, others can benefit from this market. Among these advantages are that it can be used to reduce the load during the peak of electricity consumption, the costs incurred for maintenance and operation, and uncertainty in electrical systems. In today's advanced power grids, opportunities have been created to optimize power flows, energy efficiency of power systems, as well as grid uncertainty and stability using demand response (DR) and TE. The resources used in DR and TE systems have two basic areas and are usually unevenly deployed, presenting many challenges in resource management [12]. The advantages and disadvantages of the challenges of centralized and decentralized approaches can be solved, which are presented and thoroughly discussed in [13].

Researchers in [14] have thoroughly reviewed and analyzed the latest TE technologies and their applications in today's smart power systems and reported that most of them are related to the following.

- Integration and definition of TE systems with intelligent systems
- A more comprehensive review and analysis of TE research
- Research projects and industrial exhibitions

Also, the classification of the research done in the field of TE based on network applications into the areas of (1) network transaction management (2) transaction control (3) P2P market which can identify the current challenges of TE systems and future directions.

## 2.9 Concluding Remarks

The electricity industry needs a fundamental and rapid change due to climate change, and in this, P2P trade can play an important role in different models to accompany the consumer with the producer. Also, these markets can update and advance the efforts to innovate the discussed models in the electricity industry. Therefore, in this chapter, the authors have tried to examine the concepts, challenges, and perspectives of the P2P market and energy trading and provide a correct understanding of P2P and the TGS decentralized energy system, which are generally discussed in the following sections.

In the first part of this chapter, P2P energy exchanges and TE are introduced. Recently, P2P trading has been developed as a new mechanism to strengthen the commonality between direct electricity players, which can also take responsibility and maintain the players' privacy.

There is no doubt that P2P trading has revolutionized new-generation energy management with the help of existing systems. Also, the many research works that researchers have done on P2P in the past years are indicative of this. But there is still a research gap in this field, and that is in the academic field, where the universities of the world have not yet shown much interest in this issue. In this market, a control method called TE is also provided to increase safety and make the network more efficient. There are different roles and responsibilities in this business and actors can choose and act on them. To form this market, the following structure should be considered:

- Extensive communication between actors
- Cooperation of large systems with actors
- Building trust and interaction
- Customized application and automatic business processing

The benefits of P2P are discussed and it is said that P2P is a completely new architecture in decentralized markets that allows neighboring customers to easily purchase the energy they need directly and without intermediaries. Considering that actors must be able to cope with the system complexity and coordinate energy consumption and production, P2P has been able to do this easily using TEM. The application of blockchain technology in energy was also discussed and the blockchain formation from the beginning to the time it is used in P2P was described. The existing protocols and standards in P2P business were also reviewed. Considering the increasing integration of smart power TEM systems and the risks of cyber-attacks, the impact of these attacks on P2P was presented.

## References


1. Abrishambaf, O., Lezama, F., Faria, P., & Vale, Z. (2012). Towards transactive energy systems: An analysis on current trends. *Energy Strategy Reviews, 26*, 100418.
2. Zhang, C., Wu, J., Zhou, Y., Cheng, M., & Long, C. (2018). Peer-to-peer energy trading in a microgrid. *Applied Energy, 220*, 1–12.
3. Wu, Y., Wu, Y., Cimen, H., Vasquez, J. C., & Guerrero, J. M. (2022). Towards collective energy community: Potential roles of microgrid and blockchain to go beyond P2P energy trading. *Applied Energy, 314*, 119003.
4. Javadi, M. S., Nezhad, A. E., Jordehi, A. R., Gough, M., Santos, S. F., & Catalão, J. P. (2022). Transactive energy framework in multi-carrier energy hubs: A fully decentralized model. *Energy, 238*, 121717.
5. Zhou, Y., Wu, J., Long, C., & Ming, W. (2020). State-of-the-art analysis and perspectives for peer-to-peer energy trading. *Engineering, 6*(7), 739–753.
6. Dasgupta, R., Sakzad, A., & Rudolph, C. (2021). Cyber attacks in transactive energy market-based microgrid systems. *Energies, 14*(4), 1137.
7. Zhang, C., Wu, J., Long, C., & Cheng, M. (2017). Review of existing peer-to-peer energy trading projects. *Energy Procedia, 105*, 2563–2568.
8. Said, D., Elloumi, M., & Khoukhi, L. (2022). Cyber-attack on P2P energy transaction between connected electric vehicles: A false data injection detection based machine learning model. *IEEE Access, 10*, 63640–63647.
9. Lüth, A., Zepter, J. M., del Granado, P. C., & Egging, R. (2018). Local electricity market designs for peer-to-peer trading: The role of battery flexibility. *Applied Energy, 229*, 1233–1243.
10. Soto, E. A., Bosman, L. B., Wollega, E., & Leon-Salas, W. D. (2021). Peer-to-peer energy trading: A review of the literature. *Applied Energy, 283*, 116268.
11. Guerrero, J., Chapman, A. C., & Verbič, G. (2018). Decentralized P2P energy trading under network constraints in a low-voltage network. *IEEE Transactions on Smart Grid, 10*(5), 5163–5173.
12. Siano, P., De Marco, G., Rolán, A., & Loia, V. (2019). A survey and evaluation of the potentials of distributed ledger technology for peer-to-peer transactive energy exchanges in local energy markets. *IEEE Systems Journal, 13*(3), 3454–3466.
13. Jogunola, O., et al. (2017). Comparative analysis of P2P architectures for energy trading and sharing. *Energies, 11*(1), 62.
14. Morstyn, T., Teytelboym, A., Hepburn, C., & McCulloch, M. D. (2019). Integrating P2P energy trading with probabilistic distribution locational marginal pricing. *IEEE Transactions on Smart Grid, 11*(4), 3095–3106.

15. Guerrero, J., Gebbran, D., Mhanna, S., Chapman, A. C., & Verbič, G. (2020). Towards a transactive energy system for integration of distributed energy resources: Home energy management, distributed optimal power flow, and peer-to-peer energy trading. *Renewable and Sustainable Energy Reviews, 132*, 110000.
16. Schwidtal, J. M., et al. (2022). Emerging business models in local energy markets: A systematic review of peer-to-peer, community self-consumption, and transactive energy models. In *Community self-consumption, and transactive energy models-Preprint submitted to Renewable & Sustainable Energy Reviews.*
17. Nakayama, K., Moslemi, R., & Sharma, R. (2019). Transactive energy management with blockchain smart contracts for P2P multi-settlement markets. In *2019 IEEE power & energy society innovative smart grid technologies conference (ISGT)* (pp. 1–5). IEEE.
18. Aste, T., Tasca, P., & Di Matteo, T. (2017). Blockchain technologies: The foreseeable impact on society and industry. *Computer, 50*, 18.
19. Gramoli, V. (2020). From blockchain consensus back to Byzantine consensus. *Future Generation Computer Systems, 107*, 760–769.
20. Mihaylov, M., Jurado, S., & Moffaert, K. (2014). Nrg-x-change – A novel mechanism for trading of renewable energy in smart grids. In *Proceedings of the 3rd international conference on smart grids and green IT systems* (pp. 101–106). SciTePress.
21. Cioara, T., Pop, C., Zanc, R., Anghel, I., Antal, M., & Salomie, I. (2020). Smart grid management using blockchain: Future scenarios and challenges. In *2020 19th RoEduNet conference: Networking in education and research (RoEduNet)* (pp. 1–5). IEEE.
22. Tushar, W., et al. (2021). Peer-to-peer energy systems for connected communities: A review of recent advances and emerging challenges. *Applied Energy, 282*, 116131.

# Chapter 3
# Transactive Energy and Peer-to-Peer Trading Applications in Energy Systems: An Overview

**Behzad Motallebi Azar, Hadi Mohammadian-Alirezachaei, and Rasool Kazemzadeh**

## 3.1 Introduction

### *3.1.1 Background Information*

Nowadays, energy systems rapidly evolve in structure and architecture to become more stable, flexible, and intelligent. In addition, by increasing the integration of renewable and distributed energy generation and aggravating small-scale local participants in the energy generation cycle, controlling and managing energy systems with conventional procedures will be a challenge due to operational problems (e.g., increase in the complexity of energy networks, increase in the level of uncertainties) in the near future [1]. Therefore, developments in this field are not only the basis of emerging technologies and approaches in the energy systems but also count as the transition toward green systems with more optimal and intelligent operation [2, 3]. Furthermore, increasing the complexity of the energy system's operation and communication are fundamental challenges, and facing them is inevitable. In this regard, transactive energy (TE) and peer-to-peer (P2P) trading approaches are being considered more and more to address these challenges in energy marketplace studies.

TE is defined as a mechanism for modeling and operating an energy system with a distributed information architecture to coordinate the various components of the system [4]. It aims to coordinate multi-carrier systems to dispatch energy at the distribution level, improve power quality, and encourage market actors to participate flexibly in the market [5]. P2P trading is also a set of actions done at the local level to

B. M. Azar (✉) · H. Mohammadian-Alirezachaei · R. Kazemzadeh
Faculty of Electrical Engineering, Sahand University of Technology, Tabriz, Iran
e-mail: b_motallebiazar98@sut.ac.ir; h_mohammadian99@sut.ac.ir; r.kazemzadeh@sut.ac.ir



V. Vahidinasab, B. Mohammadi-Ivatloo (eds.), *Demand-Side Peer-to-Peer Energy Trading*, Green Energy and Technology, https://doi.org/10.1007/978-3-031-35233-1_3

facilitate the exchange process [6] in which prosumers share their excess energy with others to increase market flexibility in addition to personal profit.

### *3.1.2 Challenges and Related Approaches*

Various promising approaches in the field of energy systems fully support TE and P2P trading objectives and propose suitable solutions for energy transaction problems. Attending to the regulation/policy framework of the market is one of the primary issues that the market participants should consider. In recent years, rapid growth in renewable generation on the demand side and the unbundling of the energy supply chain sprouts an opportunity for improving policy and regulatory frameworks with the aim of democratic access to energy markets [7]. Since energy is a vital commodity in modern communities, realizing its optimal sharing/allocation across the market is essential. Theoretically, markets with acceptable performance are able to find the optimal sharing/allocation, but in practice, markets face many fundamental shortcomings called market failures. Hence, the existence of sector-specific regulation may be helpful in improving the sharing/allocation in energy markets [8]. Generally, it can be understood that regulation/policy makers have the intention to move toward energy transition and mitigate the monopoly of large suppliers. These terms encourage participants to invest in renewable generation and increase their participation in local markets.

From the structural perspective, market players' correlation is accomplished in the TE platform, aiming to increase distributed energy resource penetration. Furthermore, in academic research, TE and P2P trading approaches have been studied in different aspects, such as layer dimension, market classification based on centralization level, differentiation of energy generation, sustainability of marketplace, and correlation with outdoor markets [9]. These aspects are usually determined based on regional climate, policies, and commercial factors.

In the TE approach, the primary focus is on the decentralization problems. On this basis, the structural classification of TE is divided into three major markets: fully decentralized, community-based, and network-based markets [10]. But in P2P trading, classification based on the centralization level has more attendants. Accordingly, the P2P trading markets are categorized into centralized, decentralized, and distributed markets [9]. Both the structures explained for TE and P2P trading follow similar patterns. The objectives of these market structures and their similarities are described in detail in the relevant section.

After the structural aspects of trading markets, utilizing new technologies to facilitate transactions in the approved policies and regulations, framework is necessary and inevitable. At the trading level, P2P contracts and trading management are subjects that require special attention. Therefore, various concepts and frameworks have been proposed for it. Bitcoin's emergence in 2008, which was established to create a cryptocurrency system, caused blockchain to be introduced as a secure and reliable framework for bilateral trading without needing a third party

[11, 12]. Furthermore, raising the blockchain platform in the energy markets applications made the efficiency of this platform in TE and P2P trading more visible due to transparency, tamper-proofing, and security features [13]. For instance, a P2P market, which is implemented through a decentralized blockchain-based platform, allows market players to exchange energy directly without involving a third party.

Besides technology applications, contracts also play a substantial role in energy markets. Therefore, smart contracts are employed in blockchain to perform this trading mechanism efficiently [14]. For instance, in the energy market, prosumers interact with each other to eliminate part of their demands through P2P smart contracts. Then, they interact with the upstream grid (mostly the distribution network) through the Peer-to-Grid (P2G) smart contracts (i.e., retail-based contracts) to supply the rest of their demands and perform market clearing.

Smarting techniques are a set of actions in which extensive data analysis and artificial intelligence (e.g., machine learning) are utilized to provide cognitive awareness to inanimate devices. From this viewpoint, the Internet of Things (IoT) is a concept that offers an advanced communication capability between devices (based on human preference) through the Internet for energy saving, status monitoring, and remote control [15]. For instance, in the IoT-based P2P platform, an essential part of prosumers' actions is to monitor their generations and demands and match them according to the other parties' preferences with the aim of optimal participation [16].

Improvements to the communication systems and the receiving of numerous information through smart facilities (such as sensors and smart meters) have made monitoring, handling, and accurate management of the system by a central unit for optimal decision-making a challenging work [17]. These challenges originate from various errors, faults, and time delays in receiving, processing, and transferring information. One of the promising solutions for data-driven decision-making and control techniques is machine learning algorithms [18], which have fully overcome these challenges. Through machine learning processes at the energy markets' surface, the agents continuously improve their required knowledge and experience to interact with the environment to predict actions, precept collaborative patterns, and provide solutions according to their preferences to achieve maximum efficiency [19].

### 3.1.3 Chapter Contribution and Organization

According to the aforementioned descriptions, this chapter intends to overview novel approaches in the energy trading markets, which comprise a set of challenges from research papers. Indeed, this review chapter highlights the capabilities and novelties of TE and P2P trading applications in technological and smartening aspects (e.g., blockchain technology and IoT), decision-making algorithms, along with policy, regulation, and structural issues in energy markets.

The rest of this chapter is organized as follows. Section 3.2 describes the importance of regulatory and structural issues in TE and P2P trading. Section 3.3 discusses the role of technology in TE and P2P trading. Then, the role of decision-making processes in TE and P2P trading expresses in Sect. 3.4. The conclusion and remarks are also provided in Sect. 3.5.

## 3.2 Policy, Regulation, and Structure: Importance in TE and P2P Trading

Energy markets are substantial systems utilized by policymakers, researchers, and industry to investigate techniques in order to supply/demand balance, bid in uncertain markets, and forecast prices. At the energy market level, market power is one of the determinate factors affecting market participants' benefit and competition level [20]. Hence, regulatory entities introduced different methods to mitigate market authority which can be categorized into structural and behavioral techniques. Structural techniques are utilized to modify market architecture and centralization levels. In contrast, behavioral strategies are focused on supplier behavior regardless of structure (e.g., price limitations) [21]. Therefore, it can be concluded that by performing changes in the energy market structure, it is inevitable and essential to update the market's regulatory frameworks to guarantee the dynamism and competition of energy markets. The following sub-sections explain the importance and role of energy markets' regulation and structure.

### *3.2.1 Policy and Regulation Role in Energy Markets*

Regulation in energy markets began in 1980 in a few leading countries and currently has become widespread with varying details in many countries. In this regard, policymakers' role in reforming and updating regulatory frameworks is a fact that cannot be ignored. Also, reforms and updates in market regulations have direct/indirect effects on the market participants and the long/short-term economy [22]. Comprehensive reviews of existing rules provide a starting point for ameliorating the relevant legal procedures of energy markets. For instance, authors of [23] explained the European Parliament regulations on risk-preparedness in the electricity sector crises. These regulations are intended to fulfill the Energy Union's objectives, which include energy security, solidarity, and an ambitious climate policy. Authors of [24] reviewed the implementation state of local energy markets and P2P trading based on grid tariffs in European countries. Also, authors of [25] investigated aggregators' upcoming role in implementing and operating distributed energy resources according to the key aspects of the European regulations and directives of 2019 [26, 27] in energy systems. An analysis of current regulations for self-consumption and energy sharing in societies is investigated in [28], and a general settlement framework for renewable-based communities is presented.

The regulatory framework aims to establish a relationship between renewable-based peers and the primary network [29]. In addition, market regulation is extremely complicated due to the variety of schemas, business models, prosumers types, and grid-related issues [30]. Various countries have taken steps to update their energy market policies and regulations, mainly to take advantage of RES prosumers and local energy markets. Authors of [31] proposed a review of Japan's electricity market deregulation. This study's apperceived significant policy consequences are the region's price interdependencies, indicating the difference between the regional supply/demand patterns. Therefore, this review recommends instant investment in the real-time market to simplify the interregional supply/demand balance. Another study in [32] provides a brief description of developments in the UK energy market, considering five key elements. Authors of [33] proposed an auction schema in the Brazilian energy market based on continuous structural changes in regulations with the objective of establishing a model to enhance economic efficiency, increase investment with the aim of installed capacity expansion, and guarantee market services through the competitive environment. Similarly, an outlook for identifying and understanding the main regulatory barriers of the energy sector to develop prosumer-driven business models in the Brazilian energy market is presented in [34]. In [35], a recommendation for reforming the Chinese energy market to attain a competitive market is presented, which includes implementing short-term markets, autonomous regulation, adaptability with renewable-based generations, and coordination at the provincial level. To sum up, Table 3.1 summarizes considered applications in the energy markets' regulation and policy references.

### *3.2.2 Structure Role in Energy Market*

In the last decade, a rapid expansion in the use of distributed renewable-based generations and storage devices on the consumer side caused fundamental transformations in conventional network structures. With the emergence of TE approaches, consumers' traditional structures changed and converted to prosumers. Therefore, finding a suitable structure to increase the scalability and flexibility of energy markets is quite noticeable.

This sub-section aims to describe the structural classification of TE and P2P trading. As mentioned in the previous sections, both TE and P2P trading have similar structural patterns, as presented in Figs. 3.1 and 3.2.

According to these two figures, it can be seen that the structure of the fully decentralized TE market is the same as the decentralized P2P trading market. The architecture of the community-based TE schema is similar to the centralized P2P trading market schema, and the network-based TE market model is identical to the distributed P2P trading market model. In order to better focus on structural

**Table 3.1** Summary of considered applications in the energy markets' regulation and policy references

| Reference | Year | Application | Key activity/Objective |
|---|---|---|---|
| [23] | 2019 | Regulations of risk-preparedness in the European energy sector | Energy security, solidarity, and competent climate policy |
| [24] | 2022 | Designing a framework for local electricity market | Analyzing the market status based on local trading, regulatory frameworks, and innovative grid tariff |
| [25] | 2022 | European regulatory framework in the energy transition segment | Aggregators' role in assessment, identification, and taxonomy of business models for the net-zero-emission energy transition |
| [28] | 2022 | Self-consumption and energy-sharing regulations in societies | Proposals and frameworks for energy communities and market stakeholders to improve collective self-consumption |
| [31] | 2022 | Regulation of Japan's electricity market | Investment in a real-time market to simplify regional supply/demand balance |
| [32] | 2018 | UK electricity market reformation | Analyzing five key elements in the UK energy liberalization |
| [33] | 2021 | Regulations of Energy market auction schema in Brazil | Establishing a competitive model to promote economic efficiency and competition and increase installed renewable capacity |
| [34] | 2022 | Regulatory barriers in prosumers' aggregation in the Brazilian energy sector | Identifying the regulatory challenges and developing business models by improving relevant legal frameworks |
| [35] | 2020 | Chinese energy market reformation | Developing market-based competition through short-term markets, independent regulation, and inter-provincial coordination |

approaches and not repeat analogies, we review only the structural models for TE markets.

TE markets have been introduced in various references with different approaches, depending on the grid topology and players' role [36]. Indeed, TE-based markets can be classified into three main structures according to the transaction process and the communication procedure of participants, which are explained in separate sub-sections.

#### 3.2.2.1 Fully Decentralized TE Market

Fully decentralized TE markets (Fig. 3.2a) Contain a fully P2P architecture that allows prosumers to interact with each other independently without third-party intervention and compromise on energy trading parameters [37]. In [38], a fully decentralized TE market is proposed to facilitate energy transactions, where each participant first tries to maximize its welfare based on a P2P manner and then creates

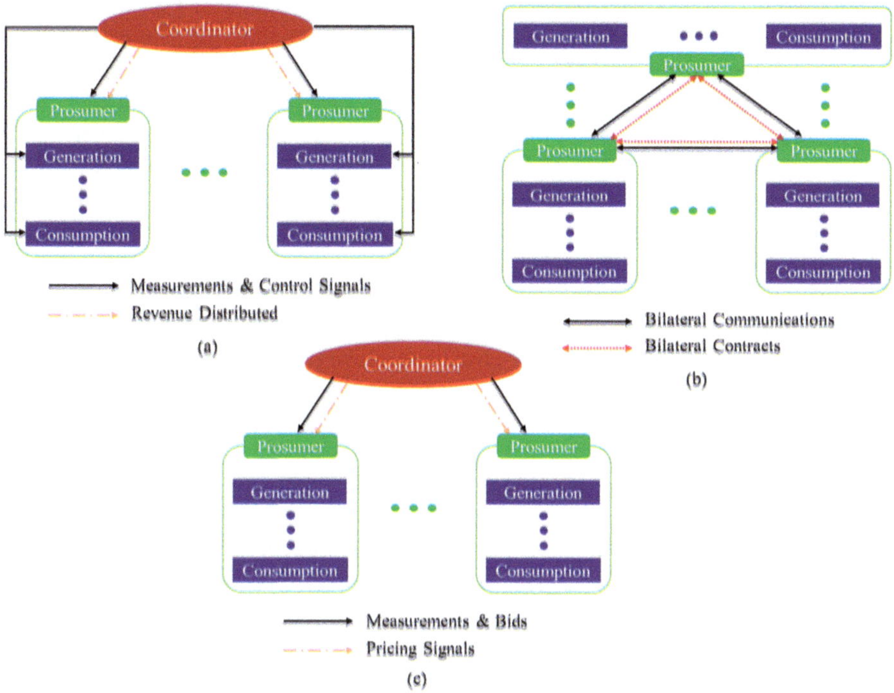


**Fig. 3.1** Structural schemas of the energy markets for P2P trading: (**a**) Centralized market, (**b**) Decentralized market, (**c**) Hybrid/distributed market [9]

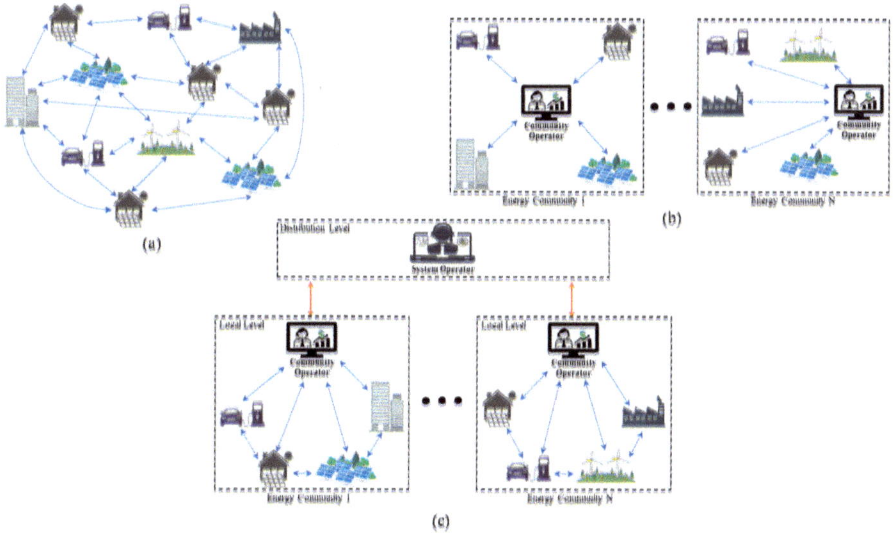


**Fig. 3.2** Structural schema of energy markets for TE: (**a**) Fully decentralized market, (**b**) Community-based market, (**c**) Network-based market [10]

a competitive environment for the energy exchange. Authors of [39] developed a fully decentralized P2P schema based on multi-bilateral economic dispatch between various market participants. In [40], a competitive TE structure containing local prosumers and retailers is developed, where local actors can interact with each other and retailers in a bilateral (P2P) manner. A TE approach for optimal management of multiple energy hubs (EHs) has been developed in [41]. This framework allows EHs to independently participate in local P2P energy markets in the first phase and then in the district market to minimize their energy costs and $CO_2$ emissions by utilizing stochastic programming. The advantages of this type of market include autonomy, empowering active participants, and buyers' complete control over their decision-making process. But in contrast, this structure also has its own disadvantages, like high investment and maintenance costs and slow convergence for final energy delivery [42].

#### 3.2.2.2 Community-Based TE Market

Community-based TE markets (Fig. 3.2b) can be applied to participants with similar interests. Unlike fully decentralized TE markets, this type of market architecture includes a community operator (CO) (or so-called coordinator) who acts as an intermediary and manages and coordinates business activities within the community. Energy exchange between indoor and outdoor entities of the communities is also managed by the CO, considering the participants' preferences. Examples of energy communities are virtual power plants (VPPs) and microgrids (MGs).

The authors of [43] developed a decentralized method to distributed energy resources' aggregation, labeled as federated power plants (FPPs). This study illustrates FPP as VPP, which allows active buyers to engage in VPP-like network transactions to improve the allocation of distributed renewable-based generations. In addition, in [44], a community and fairness-based TE market structure is proposed, which utilizes a distributed optimization technique for energy communities' operation and allows prosumers to optimize their assets actively and trade energy among themselves. A community-based transactive platform based on the multi-class management concept is presented in [45] to coordinate and facilitate prosumers' transactions beyond financial preferences. Similarly, the authors of [46] developed a community-based TE framework to facilitate prosumers' P2P trading aims to maintain the energy contracts' stability. In [47], a two-stage robust stochastic approach is provided for optimal scheduling of a community of commercial MGs considering fully renewable generations in both day-ahead and real-time manners to manage the energy trading of MGs locally and with the upstream grid. Furthermore, a novel framework for the participation of MGs in the TE platform is proposed in [48]. The main scope of the presented schema is to establish an accessible environment for microgrids' local energy trading, considering stochastic programming and information gap decision theory (IGDT) technique with risk-averse and risk-seeker strategies. The advantages of this type of market include more predictability for network operations, scalability for computing and infrastructure applications, and more compatibility with existing systems. But, the presence of a third party (i.e., a coordinator) is one of the significant disadvantages of this market structure [42].

#### 3.2.2.3 Network-Based TE Markets

The network-based TE market is aggregated form of fully decentralized and community-based TE markets. This type of market is divided into two operational layers (Fig. 3.2c) [36]. At the top layer, individual or group peers exchange energy directly in a fully decentralized manner. While at the bottom layer, energy clusters act as a community-based market companion with CO.

Network-based TE markets are most compatible with present and future systems due to their characteristics of both fully decentralized and community-based TE markets. In [49], to represent the benefits of network-based markets, a TE scheme is formulated to help reduce customers' peak electricity demand in a centralized system and facilitate interaction between local and upstream entities. The authors of [50] proposed a chance-constrained framework for MG clusters to manage energy trading to mitigate the system affiliation on the primary grid and cost-saving for the MGs. In this framework, a chance-constrained programming method is considered to model uncertainty parameters and solve the problem in realistic conditions.

The advantages of this market include a semi-independent hybrid structure controllable with CO and increasing social cooperation and resilience among participants. On the other hand, the disadvantages of this market structure are the lack of fairness and impartiality for energy sharing between participants [42].

To sum up, in a community-based market (centralized model), all market participants, who often have common interests and goals, need a primary node to monitor and settle their transactions. In fact, an institution called the community operator has been appointed to monitor this matter. Unlike this model, in the decentralized structural model, each market participant is allowed to negotiate directly with adjacent prosumers to decide their trading interests in a bilateral manner. This advantage leads to the further scalability of the market. But the network-based model (distributed trading market model) combines the above two models, leading to the division of market participants into different groups. In each of these groups, the community operator is responsible for monitoring internal and external transactions and establishing coordination between them in order to interact with the market and the system operator and conduct decentralized transactions. Moreover, a brief description of structural differences in these three models is displayed in Fig. 3.3 [51].

Finally, Table 3.2 overviews the market structures explained in this sub-section.

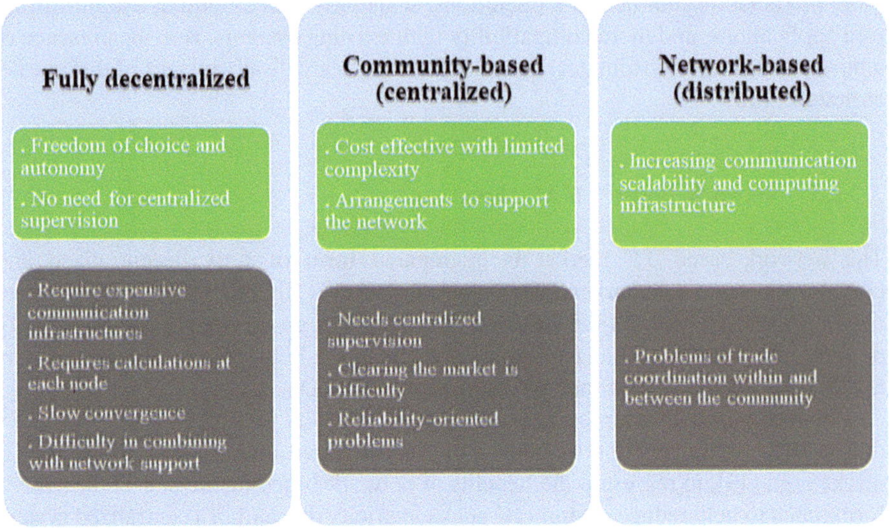


**Fig. 3.3** Structural differences of market models [51]

**Table 3.2** Overview of market structures

| Reference | Year | Application | Algorithm/Technique used |
|---|---|---|---|
| [38] | 2019 | Framework for establishing a fully decentralized market | Alternating direction method of multipliers (ADMM) |
| [39] | 2018 | P2P decentralized market structure | Multi-bilateral economic dispatch, Relaxed consensus-innovation |
| [40] | 2021 | Decentralized competitive market design | Primal-dual sub-gradient algorithm (PDSGA) |
| [41] | 2021 | TE approach for optimal energy management in local and district trading markets | Stochastic programming, holistic EH model, double auction mechanism |
| [43] | 2018 | VPP-based P2P transactions | Federated power plant (FPP) |
| [44] | 2018 | Community-based market structure | Energy collective concept |
| [45] | 2018 | P2P energy market platform | Multi-class energy management concept, ADMM, model predictive control (MPC) |
| [46] | 2020 | Framework to facilitate the P2P trading | Cooperative and non-cooperative game model, Shapely-Shubik method |
| [47] | 2020 | Optimal scheduling of commercial MGs | Two-stage robust and stochastic programming, demand response programming |
| [48] | 2021 | Model for MGs participation in the TE market | Stochastic programming, IGDT, risk-averse and risk-seeker strategies |
| [49] | 2019 | Network-based TE markets schema | Cooperative Stackelberg game, price-based coordination |
| [50] | 2020 | Framework to manage MG clusters energy trading | Chance-constrained programming |

## 3.3 Technology Role in TE and P2P Trading

Currently, traditional methods are no longer suitable applications for responding to systems development. Therefore, there is an immediate need to modernize, and new technologies and smartening techniques are emerging and flourishing. This advancement of science is clearly felt in the energy sector. In technology subjects, one newly emerging science is blockchain technology, which includes various fields. On the other hand, the acceleration in utilizing DERs, electric vehicles, storage devices, and different smart devices, as well as the prominent presence of energy prosumers, has caused smartening techniques (e.g., IoT) to become a significant part of energy systems, specifically in energy trading. New intelligent techniques and technologies in the TE and P2P trading approaches allow consumers to leave the one-dimensional state and have an active presence in the market and energy field. This means that consumers can trade their surplus energy under specific rules and on secure platforms without needing a third party or a central institution. It should be noted that blockchain technology and IoT-based smartening techniques are not the only solutions in the field of energy markets, and there is fierce competition between different technologies. But these two technologies are the most prominent ones in smart energy systems. The role of these technologies is investigated in separate sub-sections.

### *3.3.1 Blockchain-Based Technology*

Blockchain technology has confronted energy markets with various challenges in regulatory and implementation aspects. Therefore, in this subsection, we intend to introduce and review blockchain technology and its impact on transactive energy and P2P exchanges.

With the emergence of Bitcoin in 2008, a turning point in facilitating P2P transactions occurred, which included the economic aspects of power systems. Blockchain is basically an encrypted and distributed data structure that enables transactions between two anonymous parties and virtually eliminates the presence of another person or entity. This is one of the most brilliant features of blockchain. Therefore, the role of each of these market players should be clearly defined. In fact, this feature allows the network to move from a centralized to a decentralized and distributed form. Prominent features of blockchain technology can be classified as (1) Decentralization, (2) Anonymity, (3) Immutability (4) Consensus, (5) Finality (i.e., only one shared ledger is available for validating transactions) [52].

Structurally, a blockchain contains a set of blocks containing transaction data, time stamps, and encryption hash functions to connect to the previous block. This data is shared among all participating nodes in the blockchain after confirmation. The hash function guarantees the immutability of the contents of a block unless all subsequent blocks are changed.

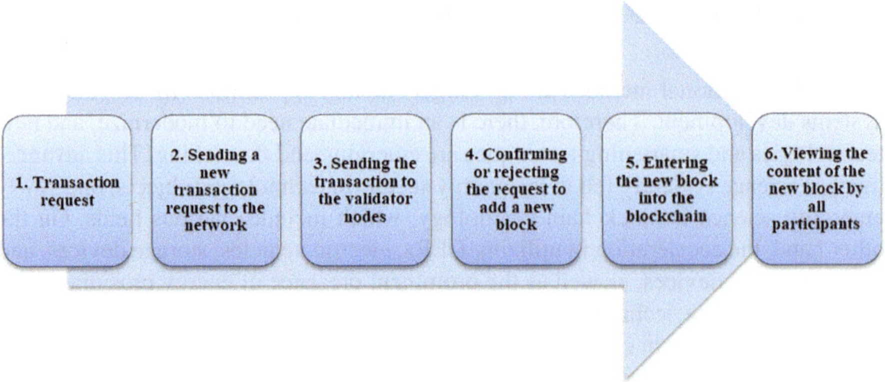


**Fig. 3.4** Procedure for adding a new transaction in the blockchain [11]

According to [11], confirming a new blockchain transaction is accomplished in six stages (as shown in Fig. 3.4): (1) Transaction request, (2) Sending a new transaction request to the network, (3) Sending the transaction to the validator nodes, (4) Confirming or rejecting the request to add a new block, (5) Entering the new block into the blockchain, (6) Viewing the content of the new block by all participants.

One part of the blockchain concept is the term "smart contracts". In fact, a smart contract is a platform to implement the logic of transactions, as well as validate transactions' accuracy. In other words, smart contracts perform automatic transactions if the market conditions are fulfilled [53]. Therefore, a smart contract programming language (e.g., solidity) is required to set market protocols to exchange tokens and energy from one peer to another.

Another part of the blockchain concept is the "consensus algorithm". In simple words, the consensus algorithm's main objective is to smooth business problems. In other words, the relevant consensus algorithm must be designed to prevent spending again or distinguish between healthy and malicious nodes [52].

In the blockchain platform, various trading mechanisms take place. But the most essential and common mechanisms are bilateral trading [54] and centralized competitive bidding [55]. In the bilateral trading technique, participants record the transaction data in the blockchain to conclude the contract. In fact, it can be noted that blockchain is only considered a distributed database in this type of mechanism. Hence, energy efficiency or social benefits cannot be imagined on this platform. On the other hand, there is a centralized competitive bidding mechanism in which the social benefit reaches its maximum value upon reaching the equilibrium point (settlement price).

Despite all the merits of the blockchain platform, there are concerns and challenges. One of the basic concerns of blockchain architecture, which is done without a third party, is double cost and fraud. Other important challenges include complexity, immutability, integration, privacy concerns, and cyber security.

According to the above content, the different applications of blockchain in the energy field, such as decentralized control of the electricity and energy grid, P2P trading, electric vehicle fleet management, and carbon emission trading, have forced researchers to do extensive work in this field. Accordingly, we are intended to depict the studies done in this field by focusing on the application of blockchain to P2P trade. If we want to describe the most fundamental studies in this field, we definitely remember the Olympic Peninsula project in 2006. A project in which consumers changed their consumption in the framework of market-based transactive control [56]. But the first known project in this matter was the Brooklyn microgrid project in New York City that was carried out in 2016 [57].

In [58], a blockchain-based platform is proposed to conduct P2P trading in MGs and guarantee the participants' security and stability. In [59], the authors analyzed the planning space of blockchain layers and their effect on market performance. Authors of [60] developed a blockchain-based P2P market in the presence of electric vehicles and tried to show the improvement of the security of financial transactions in the presence of blockchain. In [61], the authors have considered a blockchain-based platform for P2P trading, considering controllable and uncontrollable loads, PV, and energy storage, with a two-step market clearing process. In [62], the authors presented a blockchain-based management platform to optimize energy flow in an MG. In this study, the microgrid's physical constraints are formulated by optimal power flow programming combined with a bilateral mechanism, and a smart contract is implemented to execute several parts of the ADMM method as a virtual aggregator. In [63], a blockchain-based VPP energy management framework in the residential sector is designed to perform P2P energy transactions between smart houses. Similarly, in [64], a blockchain-based P2P trading framework is developed, in which two novel strategies are depicted for determining the trading preferences of households.

Many studies have been conducted in this research space, and we tried to bring suitable examples from various aspects. We can conclude that blockchain is a novel technology for energy trading studies. Generally speaking, the blockchain platform creates a safe and reliable environment for energy trading in TE and P2P trading approaches that allow market players to interact without the intervention of any central institution or intermediary. In other words, blockchain has provided a suitable solution to empower network actors to play a more active role in P2P transactive energy trading, leading to environmentally friendly DERs utilization, increase in reliability, flexibility, facilitation of market activities, and stability of electricity services.

### 3.3.2 IoT-Based Smartenig Techniques

With the emergence of smart grids, energy could be supplied cheaper, more reliable, and environmentally friendly. Also, the existence of influential smartening technologies makes microgrids' operation more optimal. One of these combinatorial technologies is IoT. The IoT is actually the fourth generation of the industrial revolution that has created new challenges for humankind. The IoT refers to an informatics system that connects different devices and elements of a system utilizing advanced information and communications technology and embedded systems (containing digital sensors, smart meters, and smart controllers) [65]. This enables devices to interact with humans and each other to achieve various goals, despite the random nature of their parameters.

Consumers who participate in the P2P market for energy sharing need to know their consumption patterns and control their equipment. This information is uploaded to the existing energy-sharing platform, and then appropriate decisions are made. In order to carry out this process, the presence of a platform such as IoT is necessary. Therefore, it helps to prepare P2P energy trading as a management method in the local energy markets [66]. In other words, IoT provides smart infrastructures to enable energy commercialization by prosumers.

In the studies carried out, authors of [67] presented an energy exchange platform based on IoT, which is related to blockchain-based monetary transactions in the form of digital currencies. The presented IoT and blockchain-based P2P platform connect participants using a web-based interface. This platform can simplify the cost-effective use of distributed generation. In [68], a blockchain-based transactional management system for smart homes is developed with IoT smartening technique aid, which considers comprehensive options for smart homes to participate in energy transactions. Therefore, smart homes can interact with the network in vertical transactions (e.g., providing demand response services) and horizontal transactions (e.g., P2P trading). Finally, to prevent private information leakage from the blockchain platform, it is suggested that this efficient blockchain system is designed according to IoT facilities, and a smart contract is developed to support the holistic TE management system. In [69], correlated smart meters form an energy IoT (EIoT) network, which enables electrical and informational transactions in the smart grid. Accordingly, customers can interact with other smart network entities in the market. This kind of perspective emerges as a multidisciplinary research subject that aims to simplify a more intelligent EIoT system. Also, plenty of existing research is focused on different aspects of blockchain systems and IoT smartening techniques, including comprehensive reviews on the application of blockchain and smartening techniques [70], privacy protection techniques [71], and blockchain-based trading systems [72]. Therefore, it can be noted that TE systems are part of smart systems that present their desired generation/consumption of energy in the form of a proposed package. In [73], a comprehensive review of utilizing the blockchain system along with IoT facilities support is presented for TE exchanges in local systems. In addition, it is necessary to create a transaction system for data/energy flows and consider human behavioral aspects to mitigate energy consumption and increase the distributed generation penetration. Therefore, authors of [74] provided an overview of interactive behaviors of buildings and electric vehicles to achieve a sustainable green TE society in three different spaces and considered key features of IoT architectures and blockchain technology as fundamental frameworks in net zero energy building.

It should be emphasized that there is no doubt about what IoT technologies will bring. Their ultimate goal is to increase the benefits of society and improve the quality of services. In the field of IoT systems, smart homes are the most considered systems. As smartening techniques move forward with digitalization, smart homes are expected to gain more popularity as an IoT framework among users due to cost-saving and energy-saving aspects [75]. But, to reach this purpose, the efficient implementation of IoT technologies with the speed and coverage of advanced wireless networks is mandatory. Hence, for a better correlation between IoT and energy systems, 5G/6G wireless systems will bring various advantages by providing more opportunities to advance and improve the current situation. In [76], a two-way trading strategy to record the random elements of the wireless 5G network from temporal, spatial, and energy pricing viewpoints with the purpose of energy saving has been addressed. In [77], a collaborative optimal operation method based on the Stackelberg game is proposed for the distribution network (DN) in the presence of a 5G communication platform. In this approach, the DN operator acts as a leader and chooses a suitable interactive price to mitigate the peak-valley difference. In [78], a 6G network infrastructure is developed for electric vehicle services based on smart contracts architecture. Another study in [79] developed a P2P trading schema for the residential sector to maximize energy sharing in a blockchain-based smart grid with 6G communications.

Finally, it can be noted that the IoT provides the possibility of integrating all types of hardware and software, platforms, services, and applications to simplify the process of interactive energy. To summarize the aforementioned articles for market smartening technologies and techniques, Table 3.3 is tabulated.

**Table 3.3** Summary of market smartening technologies and techniques in articles

| Reference | Year | Application | Algorithm/Technique used |
|---|---|---|---|
| [58] | 2021 | Sustainable MG design for P2P energy trading | Consortium blockchain platform, Fuzzy multi-objective programming, Genetic algorithm |
| [59] | 2021 | Design and evaluate the local energy market | Permissioned blockchain platform based on Hyperledger Fabric, closed-order book double auction mechanism |
| [60] | 2017 | P2P electricity market model | Consortium blockchains technology, double auction mechanism |
| [61] | 2019 | An optimization platform for P2P trading in crowdsourced energy systems | Two-phase near real-time operation algorithm, blockchain platform based on Hyperledger Fabric |
| [62] | 2020 | P2P energy trading optimization in an MG | Private blockchain platform, optimal Power Flow, bilateral trading mechanism, ADMM |
| [63] | 2021 | Energy management platform for residential users | Blockchain-based VPP, distributed optimization algorithm |
| [64] | 2021 | Decentralized P2P energy trading in the residential sector | Permissioned blockchain platform, bilateral trading mechanism |

(continued)

**Table 3.3** (continued)

| Reference | Year | Application | Algorithm/Technique used |
|---|---|---|---|
| [67] | 2020 | IoT-based P2P trading platform | Blockchain platform based on Ethereum, interactive web-based user interface |
| [68] | 2021 | IoT-based holistic TE system for smart homes | Two dimensions TE algorithm, privacy-preserving distributed algorithm, blockchain platform |
| [69] | 2018 | Internet of energy architecture for TE systems | Multitier communication architecture, inter-customer price function, demand response programming |
| [72] | 2020 | Optimal TE platform for P2P trading | Multilayer blockchain-based TE platform |
| [76] | 2017 | Two-way trading strategy for 5G communications | Lyapunov optimization method |
| [77] | 2021 | DN optimization in the presence of 5G networks | Stackelberg game approach |
| [78] | 2022 | Management of electric vehicle services with 6G networks | Smart-contract-based secure edge computing, collaborative computing resource allocation algorithm |
| [79] | 2021 | P2P trading schema for the residential sector with 6G networks | Q-learning RL algorithm, blockchain platform based on Ethereum |

## 3.4 Decision-Making Role in TE and P2P Trading

As noted earlier, decision-making techniques can be considered one of the leading indicators of modern smart systems and a solution to overcome energy systems' challenges. Machine learning methods are one of the most promising solutions presented in this regard. The basis of machine learning returns to the interest of scientific association in the 50 and 60 decades in replicating human learning through computer programs. From this viewpoint, machine learning extracts knowledge from data and then utilizes it for anticipation, producing new records, and decision-making as well [80]. Besides, using agent-based simulations is a reliable option for a better understanding of energy market dynamics, which prepares an opportunity to better forecast and control the energy market in a more efficient route [81]. Machine learning algorithms are generally categorized into supervised learning (SL), unsupervised learning (USL), and reinforcement learning (RL) [82]. These algorithms are described in the sub-sections as follows:

### *3.4.1 Supervised and Unsupervised Learning*

Supervised learning utilizes training data which includes inputs and desired outputs. In such a field, a so-called supervisor provides information to the learner to predict outputs associated with new inputs. In this type of learning, we deal with two types of variables. The first type is called independent variables (including one or more variables), which are used to predict another variable. The second type is dependent variables (also called target or output variables), and their values are predicted with the help of SL algorithms [83]. To this end, a function containing inputs (independent variables) should be created that generates the desired output (dependent variable). The process of finding this function is called the training process, which is applied based on existing data (data with known independent and dependent variables, e.g., the energy purchasing history) and continues until the required accuracy is reached [81].

Therefore, an efficient SL model can properly forecast the output of inputs not part of the primary training data. SL can be utilized for classification and regression. In the classification, the model should use its observations to specify the placement of the new observations category. This category forecasts outputs with discrete values [84, 85]. In the regression, the model must comprehend the relations between variables to approximate the output and anticipate outputs with continuous values [86, 87]. Accordingly, a number of the commonly used SL algorithms are support vector machines (SVM) [88], linear regression [89], and artificial neural networks (ANN) [90].

In unsupervised learning, unlike SL, there is no specific data in advance, and the learner must look for a particular structure in the data. In this type of learning, we do not have a target variable, and the algorithm's output is uncertain. It means that, in USL, the data are unlabeled, and the model must explore correlations and relations utilizing extant data analysis [91]. The most commonly used algorithms in USL are the Bayesian classifier [92] and K-means clustering [93]. USL is not the appropriate case when labeled data is accessible. Hence, it is less common in agent-based simulations. However, USL is a powerful tool where there is no labeled data.

It must be noted that the number of research studies utilizing SL and USL algorithms, specifically in energy markets, is much lower than in other study fields. We imply some of these studies in the following sentences.

Authors of [94] have proposed a model based on supervised learning algorithms for decision support of strategic participation of prosumers in the trading market considering risk management. In [95], an electricity price forecasting technique using ANNs is developed in an agent-based electricity market name PowerACE. Authors of [96] presented an SVM-based approach for market participants to establish decision support and bid suggestion based on the market price prediction.

A novel strategy for bilateral negotiations in the electricity market based on unsupervised learning (i.e., Bayesian classifier) is developed in [97]. This strategy enables each participant to estimate the electricity prices of their opponents. In [98], a management approach for MG operation is presented using K-means clustering, where the electricity market agent is responsible for MG management as an aggregator. In [99], an optimization formulation for the fair allocation of distributed energy generations in a local market is presented based on an unsupervised clustering method and truthful auction.

### *3.4.2 Reinforcement Learning*

Reinforcement learning is another branch of artificial intelligence systems and the most prominent model of machine learning algorithms, which is focused on performing consecutive actions of agents in an uncertain interactive environment and learning from feedback. Therefore, RL can play a significant role in overcoming challenges regarding increasing complexity and optimal decision-making [100].

The structure of RL methods can be classified into two major parts; model-based and model-free. Model-based methods refer to RL techniques that precisely appraise and update a system model online in which agents predict their actions based on collected information from this estimated environment [101]. In the model-based environment, iterative linear quadratic regulator, Thompson sampling, and probabilistic inference for learning control are the most considered algorithms. The iterative linear quadratic regulator is a non-linear model for calculating the optimal trajectory of states in an iterative manner [102]. Thompson sampling is a multi-armed bandit heuristic algorithm for choosing actions in an iterative problem to maximize the expected reward with respect to a randomly drawn belief [103]. Probabilistic inference for control learning is a dynamic decision-making method that predicts the system's future uncertain state by relying on the Gaussian process to estimate the optimal value [104].

In model-free methods, agents interact with the environment without estimating the system model and predict their actions through trial-and-error techniques [105]. In this model, agents are rewarded for their updates. By maximizing these accumulative rewards, agents attain the optimal operational strategy.

Specifically in energy markets, the model-free methods can be considered the most notable RL algorithm due to their fast and efficient operation with less complexity compared to the model-based methods. Model-free algorithms are divided into two policy-gradient (PG) and value-based parts. In the PG method, the aim is to learn the best policy for performing the best actions to maximize reward, while in the value-based method, the objective is to find the optimal value function [2]. Furthermore, the value-based nature of the RL algorithm aids it in being compatible with real-time observations and acting desirably in dynamic environments.

In the PG environment, Actor-Critic, trust region, stochastic PG, and deterministic PG are the most considered algorithms. Actor-Critic is a simultaneous method for learning both policy and value functions. In this algorithm, the policy function is counted as the actor and proposes actions given a state, and the value function is counted as the critic and evaluates actions taken by the actor [106]. Trust region is utilized for problems with extensive parameter updates in the policy function. This method avoids numerous policy changes in each iteration [107]. Stochastic and deterministic PG algorithms are other methods for integrating policy in the state and action spaces. In stochastic PG algorithms, the probability of each action in each state is determined based on specific policy [108]. But, a deterministic PG algorithm determines which action should be taken in each state [109].

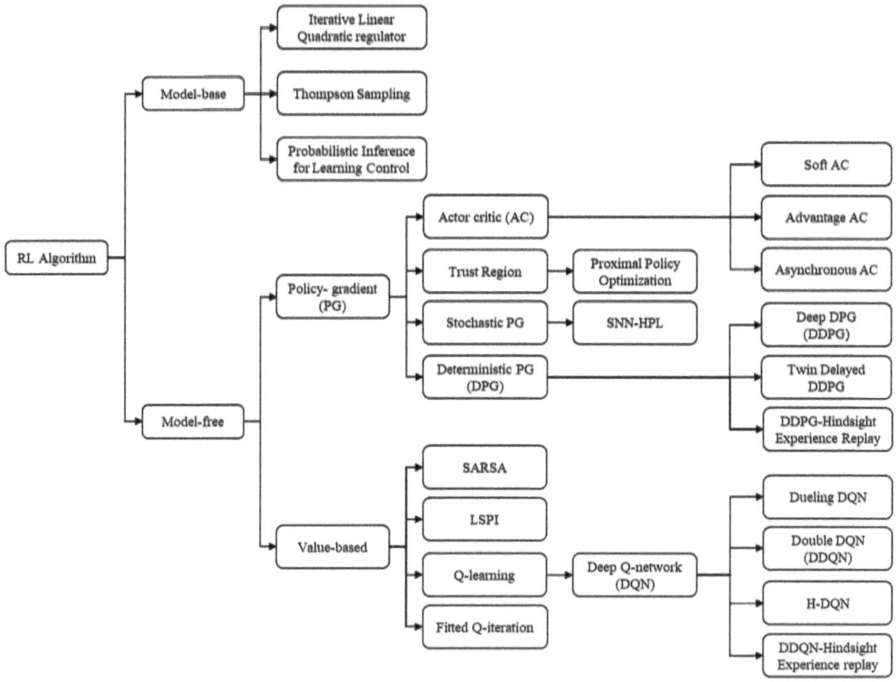


**Fig. 3.5** General classification of RL algorithms [2, 105]

SARSA, Q-learning Least-Squares Policy Iteration (LSPI) and fitted Q-iteration are the most considered algorithms in the value-based environment. SARSA algorithm is the acronym of "State, Action, Reward, State´, Action" tuple and is an on-policy learning technique designed to take actions performed by the current policies [110]. Unlike SARSA, Q-learning is an off-policy learning method designed to learn the value of an action in a particular state [111]. LSPI is another off-policy method that utilizes the load-store queue outcome to create an approximate policy iteration algorithm [112]. Fitted Q-iteration is a batch-learning technique to approximate the rewards of the Q-learning problem by iteratively extending the optimization horizon [113]. The general classification of explained RL algorithms is summarized in Fig. 3.5. The last columns of each algorithm refer to their utilized deep RL techniques.

In the RL, decision-making problems are modeled by Markov decision processes (MDP), which is a mathematical model of sequential decision-making utilized to simulate strategies and stochastic rewards. In the MDP, agents can be trained in an environment with Markov characteristics [110]. Also, it is worth noting that the whole RL methods need a development environment. Recurrent neural network (RNN), long short-term memory (LSTM), and convolutional neural network (CNN) algorithms are the most common development environments [19].

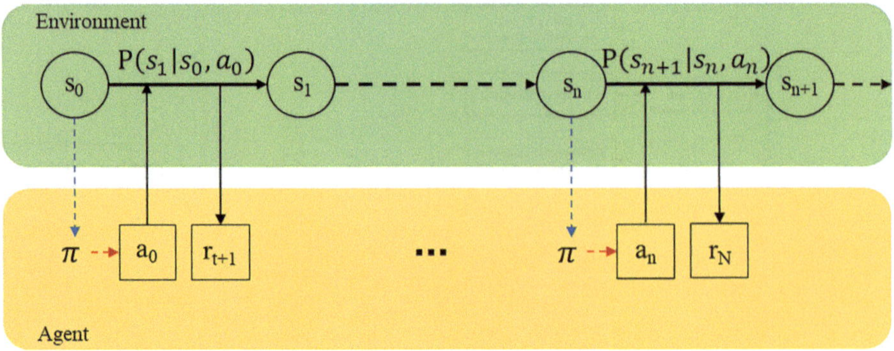


**Fig. 3.6** Procedure of MDP [99]

The procedure of MDP is demonstrated in Fig. 3.6. The MDP problem begins by placing the initial state $s_0 \in S$. At any time $t = \{0, 1, \ldots, n\}$ according to the state $s_n \in S$, the agent selects the action $a_n \in A$ and obtaines the reward $r(s_n, a_n)$. Then, the next state $s_{n+1}$ is constructed from the transition probability function $P(s_{n+1}| s_n, a_n)$. For each agent, the policy $\pi(a| s) \in \Delta(A)$ is a mapping of state ($s$) for distribution in the action space ($A$). This policy provides a principle about which action must be adopted according to the state ($s$) particular situation. In the P2P trading environment, action ($A$) represents the actors' participation, and the reward function ($R$) also demonstrates the received incentive through performing specific actions [19, 105, 114].

According to the explanations provided, it can be concluded that agents of prosumers significantly need the learning ability to make more accurate decisions. For this scope, RL decision-making approaches can be considered one of the most influential factors in the operation of TE and P2P energy markets.

In [115], a P2P trading schema based on blockchain technology and machine learning algorithms is presented to provide real-time support, day-ahead control, and generation planning of distributed renewable-based sources. This proposed schema includes three modules for P2P exchange, data analysis, and predictions. The trading module uses a smart contract on the Hyperledger Fabric blockchain with the byzantine fault tolerance (BFT) consensus algorithm. With the help of data analysis and prediction modules, the agents make effective decisions in order to minimize energy costs. A P2P trading model using reinforcement learning methods is developed in [110], in which the rule of electrical storage systems of prosumers in electricity transactions is discussed. In this model, a SARSA RL algorithm has been utilized to analyze and solve problems related to prosumers' decision-making. In [116], an automatic P2P transactive method based on RL algorithms utilizing the short/long delay reward (LSTDR) technique is proposed to maximize the individual profit of prosumers. Also, a deep Q-Network (DQN) has been utilized as a training model to analyze time-dependent information more effectively. Authors of [117] developed a bi-level TE model for optimal real-time management of networked MGs to minimize operational costs based on DDQN RL algorithms. In this approach, the DN operator as a controller is at the upper level, and the MGs are also at the bottom level. Also, to improve the performance of both levels, an interactive technique based on the Single-leader-Multi-follower Stackelberg game is presented, in which the DN operator is the leader and the MGs are the followers. A transactive schema is presented in [118] for managing the prosumers' transactions using RL-based algorithms. In this framework, a local energy market and community energy storage are designed and modeled that control prosumers' actions in a decentralized manner. Authors of [119] proposed a TE concept for optimal bidding strategy based on a soft actor-critic learning method in an active DN in which actors participate in a local market and submit their bids to the TE market for day-ahead scheduling. In [120], the behavioral pattern of prosumers in a local P2P market is also modeled to facilitate energy trading at the distribution level. Authors of [121] introduced an independent local transactive platform incorporating multi-agent RL algorithms and double-auction mechanisms. In this regard, participants specify their pricing signals and manage their decisions through market interactions. In [122], a hierarchical deep RL-based transactive framework is developed for a society of residential buildings with a two-stage learning method. A deep RL-based algorithm is utilized in both stages for appliance scheduling and price determination. Similarly, authors of [123] have proposed a TE model based on a multi-agent RL approach and an attention mechanism for the microgrids' optimal learning. Each MG is modeled as an independent agent in this framework and learns to manage its transactions in cooperation with other agents. Authors of [124] have explored a model-free prosumer-centric coordinated energy trading mechanism for a local market to address the practical restrictions of model-based system-centric models through a multi-agent deep RL method. In [125], a P2P market framework using double auctions and model-based RL is developed, in which the prosumers participate in the market through their own bidding history, and payoffs are determined based on bids and auction mechanisms. Also, the authors of [126] presented a P2P trading model for the household sector and applied multi-agent deep RL algorithms with a decentralized transactive schema to attain prosumers' optimal participation. For simplification, described studies related to decision-making approaches are tabulated in Table 3.4 as follows:

**Table 3.4** Studies related to decision-making approaches

| Reference | Year | Application | Algorithm/Technique used |
|---|---|---|---|
| [94] | 2019 | Market actors' participation risks | Internal and sectorial forecasting approaches, ANN |
| [95] | 2021 | Electricity price forecasting in an agent-based market | ANN |
| [96] | 2016 | Market price forecasting for decision support and bid suggestions | ANN, SVM |
| [97] | 2020 | Bilateral trading in the electricity market | Bayesian classifier |
| [98] | 2021 | MG management by market agents | K-means clustering |
| [99] | 2019 | Local energy market optimization | Expectation and maximization (EM) learning algorithm, truthful auction mechanism |

(continued)

**Table 3.4** (continued)

| Reference | Year | Application | Algorithm/Technique used |
|---|---|---|---|
| [110] | 2020 | P2P electricity transaction model | SARSA RL algorithm |
| [115] | 2021 | P2P trading platform for real-time support, day-ahead control, and renewable generation planning | Smart contract based on Hyperledger Fabric, deep learning with RNN, LSTM, Bi-directional LSTM |
| [116] | 2020 | Automatic P2P trading model | DQN RL algorithm with LSTDR |
| [117] | 2021 | TE framework for microgrids' real-time management | DDQN RL algorithms, single-leader-multi-follower Stackelberg game |
| [118] | 2021 | TE management framework to control prosumers' actions | RL-based trading algorithm |
| [119] | 2022 | TE concept for optimal bidding strategy | Soft AC RL algorithm |
| [120] | 2018 | Behavioral pattern to facilitate P2P trading | DQN RL algorithm |
| [121] | 2022 | Autonomous local energy exchange (ALEX) framework | Multi-agent Q-learning RL algorithm, Double auction mechanism |
| [122] | 2022 | Real-time trading in the community market | Multi-agent H-DQN RL algorithm |
| [123] | 2022 | Optimal decision-making for energy trading among MGs | Multi-agent Q-learning RL algorithm with an attention mechanism |
| [124] | 2022 | Coordination of local energy markets | Multi-actor-attention-critic RL algorithm |
| [125] | 2022 | P2P energy trading framework | Model-based RL algorithm, Double auction mechanism |
| [126] | 2022 | Optimal energy trading and demand-side management strategies in a P2P market | Multi-agent DQN RL algorithm, multi-agent DDPG RL algorithm with independent Q-learning training structure |

## 3.5 Conclusion and Remarks

Despite the progress made in energy trading and marketing, many challenges and barriers still need to be investigated until appropriate solutions are provided. These challenges and barriers can be divided into two general categories: technical and legal. Therefore, this chapter provides an overview of TE and P2P trading applications in the surface of energy systems, comprising challenges in policy/regulation, structure, technology, and decision-making approaches.

Section two briefly describes the energy market regulations and structures. Accordingly, it can be concluded that regulatory and structural frameworks are correlated, and any change in the market structure requires new rules or updating existing regulations. Hence, there is a need to move toward more practical and comprehensive studies based on regulatory/structural correlations. The third section explains the role of novel technologies and smartening techniques. In this regard, blockchain technology and the IoT smartening technique are the most prominent subjects intended for academic research. Currently, the blockchain platform is a commonly used technology in energy market planning, which creates a reliable environment for energy trading with a variety of access levels without the intervention of any central institution or intermediary. In other words, blockchain has provided a suitable solution to empower network actors to play a more active role in P2P transactional energy trading. Also, the IoT framework is a brilliant informatic system as the fourth generation of the industrial revolution, with capabilities to link various objects and elements to serve humankind optimally. The results obtained from combining IoT and blockchain infrastructure in multiple literatures indicate an operational improvement in TE and P2P trading systems. Also, adding 5G/6G communicational networks in IoT-based systems is a novel research area in TE and P2P trading studies. The fourth section also addresses the importance of decision-making approaches in energy markets. Machine learning methods are one of the most promising agent-based solutions presented in this regard. Among the machine learning methods, RL algorithms are the most considered ones. Currently, RL algorithms are commonly used in optimal operation problems, especially participating approaches in P2P energy markets, due to fast response and a wide range of applications. However, to prove RL algorithms' efficiency, there is a need to benchmark these algorithms in more practical fields, especially in comparison with distributed algorithms (e.g., ADMM).

As a future outlook, attending to market flexibility factors in energy trading and transitioning from passive centralized to advanced decentralized systems in the presence of advanced communication systems are vital points that challenge energy markets theoretically and practically.

## References


1. Gungor, V. C., Sahin, D., Kocak, T., Ergut, S., Buccella, C., Cecati, C., & Hancke, G. P. (2011). Smart grid technologies: Communication technologies and standards. *IEEE Transactions on Industrial Informatics, 7*(4), 529–539.
2. Jogunola, O., et al. (2021). Consensus algorithms and deep reinforcement learning in energy market: A review. *IEEE Internet of Things Journal, 8*(6), 4211–4227.
3. Azar, B. M., Kazemzadeh, R., & Baherifard, M. A. (2020). Energy hub: Modeling and technology-A review. In *2020 28th Iranian conference on electrical engineering (ICEE)* (pp. 1–6). IEEE.
4. Zou, Y., Xu, Y., Feng, X., Naayagi, R. T., & Soong, B. H. (2022). Transactive energy systems in active distribution networks: a comprehensive review. *CSEE Journal of Power and Energy Systems, 8*(5), 1302–1317.
5. Li, Z. (2017). *Transactive energy system.* Honours thesis, Murdoch University.
6. Dorahaki, S., Rashidinejad, M., Fatemi Ardestani, S. F., Abdollahi, A., & Salehizadeh, M. R. (2021). A peer-to-peer energy trading market model based on time-driven prospect theory in a smart and sustainable energy community. *Sustainable Energy, Grids and Networks, 28*, 100542.
7. Moura, R., & Brito, M. C. (2019). Prosumer aggregation policies, country experience and business models. *Energy Policy, 132*, 820–830.

8. Mulder, M. (2020). *Regulation of energy markets: Economic mechanisms and policy evaluation*. Springer Nature Textbook.
9. Zhou, Y., Wu, J., Long, C., & Ming, W. (2020). State-of-the-art analysis and perspectives for peer-to-peer energy trading. *Engineering, 6*(7), 739–753. ISSN 2095-8099.
10. Adeyemi, A., Yan, M., Shahidehpour, M., Bahramirad, S., & Paaso, A. (2020). Transactive energy markets for managing energy exchanges in power distribution systems. *The Electricity Journal, 33*(9), 106868.
11. Di Silvestre, M. L., Gallo, P., Guerrero, J. M., Musca, R., Sanseverino, E. R., Sciumè, G., Vásquez, J. C., & Zizzo, G. (2020). Blockchain for power systems: Current trends and future applications. *Renewable and Sustainable Energy Reviews, 119*, 109585.
12. Liu, Y., Yu, F. R., Li, X., Ji, H., & Leung, V. C. (2020). Blockchain and machine learning for communications and networking systems. In *IEEE communications surveys & tutorials*. IEEE.
13. Andoni, M., Robu, V., Flynn, D., Abram, S., Geach, D., Jenkins, D., McCallum, P., & Peacock, A. (2019). Blockchain technology in the energy sector: A systematic review of challenges and opportunities. *Renewable and Sustainable Energy Reviews, 100*, 143–174.
14. Solanki, B., et al. (2021). Blockchain-based decentralized hybrid P2P energy trading. In *2021 9th IEEE international conference on power systems (ICPS)* (pp. 1–5). IEEE.
15. Tushar, W., Wijerathne, N., Li, W. T., Yuen, C., Poor, H., Saha, T., & Wood, K. (2018). Internet of things for green building management: Disruptive innovations through low-cost sensor technology and artificial intelligence. *IEEE Signal Processing Magazine, 35*(5), 100–110.
16. Al-Ali, A. R., Zualkernan, I. A., Rashid, M., Gupta, R., & Alikarar, M. (2017). A smart home energy management system using IoT and big data analytics approach. *IEEE Transactions on Consumer Electronics, 63*(4), 426–434.
17. Wilcox, T., Jin, N., Flach, P., & Thumim, J. (2019). A big data platform for smart meter data analytics. *Computers in Industry, 105*, 250–259.
18. Ray, S. (2019). A quick review of machine learning algorithms. In *2019 International conference on machine learning, big data, cloud and parallel computing (COMITCon)* (pp. 35–39). IEEE.
19. Ravichandiran, S. (2018). *Hands-on reinforcement learning with python: Master reinforcement and deep reinforcement learning using OpenAI Gym and TensorFlow*. Packt Publishing.
20. Banovac, E., Glavic, M., & Tesnjak, S. (2009). Establishing an efficient regulatory mechanism prerequisite for successful energy activities regulation. *Energy, 34*, 178–189.
21. Helman, U. (2006). Market power monitoring and mitigation in the US wholesale power markets. *Energy, 31*, 877–904.
22. Selçuk, I., & Köktaş, A. (2016). Energy market regulations and productivity: An examination on OECD countries between the years of 1975–2007. *Sosyoekonomi Journal, 24*, 243–261.
23. European Commission. (2019, June 14). Regulation (EU) 2019/941 of the European Parliament and of the Council of 5 June 2019 on risk-preparedness in the electricity sector and repealing Directive. *Official Journal of the European Union*. [Online]. Available: http://data.europa.eu/eli/reg/2019/941/oj
24. Maldet, M., Huglen Revheim, F., Schwabeneder, D., Lettner, G., del Granado, P. C., Saif, A., Löschenbrand, M., & Khadem, S. (2022). Trends in local electricity market design: Regulatory barriers and the role of grid tariffs. *Journal of Cleaner Production, 358*, 131805.
25. Kerscher, S., & Arboleya, P. (2022). The key role of aggregators in the energy transition under the latest European regulatory framework. *International Journal of Electrical Power & Energy Systems, 134*, 107361.
26. European Commission. (2019, June 14). Directive (EU) 2019/944 of the European Parliament and of the Council of 5 June 2019 on common rules for the internal market for electricity and amending directive. *Official Journal of the European Union*. [Online]. Available: http://data.europa.eu/eli/dir/2019/944/oj
27. European Commission. (2019, June 14). Regulation (EU) 2019/943 of the European Parliament and of the Council of 5 June 2019 on the internal market for electricity. *Official Journal of the European Union*. [Online]. Available: http://data.europa.eu/eli/reg/2019/943/oj

28. Mello, J., Villar, J., & Saraiva, J. (2022). Conciliating the settlement of local energy markets with self-consumption regulations. Available at SSRN: http://dx.doi.org/10.2139/ssrn.4097357
29. Parag, Y., & Sovacool, B. (2016). Electricity market design for the prosumer era. *Nature Energy, 1*, 1–6.
30. Heffron, R. J., & Talus, K. (2016). The evolution of energy law and energy jurisprudence: Insights for energy analysts and researchers. *Energy Research & Social Science, 19*, 1–10.
31. Suliman, M. S., & Farzaneh, H. (2022). Econometric analysis of pricing and energy policy regulations in Japan electric power exchange spot market. *Cleaner Engineering and Technology, 9*, 100523.
32. Grubb, M., & Newbery, D. (2018). UK electricity market reform and the energy transition: Emerging lessons. *The Energy Journal, 39*(6), 2–25.
33. Tolmasquim, M. T., Correia, T. B., Addas Porto, N., & Kruger, W. (2021). Electricity market design and renewable energy auctions: The case of Brazil. *Energy Policy, 158*, 112558.
34. Botelho, D. F., de Oliveira, L. W., Dias, B. H., Soares, T. A., & Moraes, C. A. (2022). Prosumer integration into the Brazilian energy sector: An overview of innovative business models and regulatory challenges. *Energy Policy, 161*, 112735.
35. Guo, H., Davidson, M. R., Chen, Q., Zhang, D., Jiang, N., Xia, Q., Kang, C., & Zhang, X. (2020). Power market reform in China: Motivations, progress, and recommendations. *Energy Policy, 145*, 111717.
36. Sousa, T., Soares, T., Pinson, P., Moret, F., Baroche, T., & Sorin, E. (2019). Peer-to-peer and community-based markets: A comprehensive review. *Renewable and Sustainable Energy Reviews, 104*, 367–378.
37. Tushar, W., Saha, T. K., Yuen, C., Smith, D., & Poor, H. V. (2020). Peer-to-peer trading in electricity networks: An overview. *IEEE Transactions on Smart Grid, 11*(4), 3185–3200.
38. Yang, J., Dong, Z., Chen, G., Wen, F., & Li, C. (2019). A fully decentralized distribution market mechanism using ADMM. In *IEEE power & energy society general meeting (PESGM)* (pp. 1–5). IEEE.
39. Sorin, E., Bobo, L., & Pinson, P. (2018). Consensus-based approach to peer-to-peer electricity markets with product differentiation. *IEEE Transactions on Power Systems, 34*(2), 994–1004.
40. Mehdinejad, M., Shayanfar, H., & Mohammadi-Ivatloo, B. (2022). Peer-to-peer decentralized energy trading framework for retailers and prosumers. *Applied Energy, 308*, 118310.
41. Khorasany, M., Najafi-Ghalelou, A., Razzaghi, R., & Mohammadi-Ivatloo, B. (2021). Transactive energy framework for optimal energy management of multi-carrier energy hubs under local electrical, thermal, and cooling market constraints. *International Journal of Electrical Power & Energy Systems, 129*(106803). ISSN 0142-0615.
42. Huang, Q., Amin, W., Umer, K., Gooi, H. B., Eddy, F. Y. S., Afzal, M., Shahzadi, M., Aman Khan, A., & Ahmad, S. A. (2021). A review of transactive energy systems: Concept and implementation. *Energy Reports, 7*, 7804–7824.
43. Morstyn, T., Farrell, N., Darby, S. J., & McCulloch, M. D. (2018). Using peer-to-peer energy trading platforms to incentivize prosumers to form federated power plants. *Nature Energy, 3*(2), 94–101.
44. Moret, F., & Pinson, P. (2018). Energy collectives: A community and fairness based approach to future electricity markets. *IEEE Transactions on Power Systems, 34*(5), 3994–4004.
45. Morstyn, T., & McCulloch, M. D. (2018). Multi-class energy management for peer-to-peer energy trading driven by prosumer preferences. *IEEE Transactions on Power Systems, 34*(5), 4005–4014.
46. Amin, W., Huang, Q., Afzal, M., Khan, A. A., Umer, K., & Ahmed, S. A. (2020). A converging non-cooperative & cooperative game theory approach for stabilizing peer-to-peer electricity trading. *Electric Power Systems Research, 183*, 106278.
47. Daneshvar, M., Mohammadi-Ivatloo, B., Zare, K., & Asadi, S. (2020). Two-stage robust stochastic model scheduling for transactive energy based renewable microgrids. *IEEE Transactions on Industrial Informatics, 16*(11), 6857–6867.

48. Daneshvar, M., Mohammadi-Ivatloo, B., Zare, K., Asadi, S., & Anvari-Moghaddam, A. (2021). A novel operational model for interconnected microgrids participation in transactive energy market: A hybrid IGDT/stochastic approach. *IEEE Transactions on Industrial Informatics, 17*(6), 4025–4035.
49. Tushar, W., Saha, T. K., Yuen, C., Morstyn, T., Poor, H. V., & Bean, R. (2019). Grid influenced peer-to-peer energy trading. *IEEE Transactions on Smart Grid, 11*(2), 1407–1418.
50. Daneshvar, M., Mohammadi-Ivatloo, B., Asadi, S., Anvari-Moghaddam, A., Rasouli, M., Abapour, M., & Gharehpetian, G. B. (2020). Chance-constrained models for transactive energy management of interconnected microgrid clusters. *Journal of Cleaner Production, 271*, 122177.
51. Nizami, S., Tushar, W., Hossain, M. J., Yuen, C., Saha, T., & Poor, H. V. (2022). Transactive energy for low voltage residential networks: A review. *Applied Energy, 323*, 119556.
52. Han, D., Zhang, C., Ping, J., & Yan, Z. (2020). Smart contract architecture for decentralized energy trading and management based on blockchains. *Energy, 199*, 117417.
53. Karamitsos, I., Papadaki, M., & Al Barghuthi, N. B. (2018). Design of the blockchain smart contract: A use case for real estate. *Journal of Information Security, 9*(3), 177.
54. Hahn, A., Singh, R., Liu, C. C., & Chen, S. (2017). Smart contract-based campus demonstration of decentralized transactive energy auctions. In *IEEE power & energy society innovative smart grid technologies conference (ISGT)* (pp. 1–5). IEEE.
55. Thomas, L., Long, C., Burnap, P., Wu J., & Jenkins, N. (2017). Automation of the supplier role in the gb power system using blockchain-based smart contracts. *CIRED – Open Access Proceedings Journal, 1*, 2619–2623.
56. Hammerstrom, D. J., Ambrosio, R., Carlon, T. A., DeSteese, J. G., Horst, G. R., Kajfasz, R., et al. (2008). *Pacific northwest gridwise™ testbed demonstration projects; Part I.* Olympic Peninsula Project (No. PNNL-17167). Pacific Northwest National Lab (PNNL).
57. Mengelkamp, E., Gärttner, J., Rock, K., Kessler, S., Orsini, L., & Weinhardt, C. (2018). Designing microgrid energy markets: A case study: The Brooklyn Microgrid. *Applied Energy, 210*, 870–880.
58. Tsao, Y. C., & Thanh, V. V. (2021). Toward sustainable microgrids with blockchain technology-based peer-to-peer energy trading mechanism: A fuzzy meta-heuristic approach. *Renewable and Sustainable Energy Reviews, 136*, 110452.
59. Christidis, K., Sikeridis, D., Wang, Y., & Devetsikiotis, M. (2021). A framework for designing and evaluating realistic blockchain-based local energy markets. *Applied Energy, 281*, 115963.
60. Kang, J., Yu, R., Huang, X., Maharjan, S., Zhang, Y., & Hossain, E. (2017). Enabling localized peer-to-peer electricity trading among plug-in hybrid electric vehicles using consortium blockchains. *IEEE Transactions on Industrial Informatics, 13*(6), 3154–3164.
61. Wang, S., Taha, A. F., Wang, J., Kvaternik, K., & Hahn, A. (2019). Energy crowdsourcing and peer-to-peer energy trading in blockchain-enabled smart grids. *IEEE Transactions on Systems, Man, and Cybernetics: Systems, 49*(8), 1612–1623.
62. Leeuwen, G. V., AlSkaif, T., Gibescu, M., & Sark, W. V. (2020). An integrated blockchain-based energy management platform with bilateral trading for microgrid communities. *Applied Energy, 263*, 114613.
63. Yang, Q., Wang, H., Wang, T., Zhang, S., Wu, X., & Wang, H. (2021). Blockchain-based decentralized energy management platform for residential distributed energy resources in a virtual power plant. *Applied Energy, 294*, 117026.
64. AlSkaif, T., Crespo-Vazquez, J. L., Sekuloski, M., van Leeuwen, G., & Catalão, J. P. (2021). Blockchain-based fully peer-to-peer energy trading strategies for residential energy systems. *IEEE Transactions on Industrial Informatics, 18*(1), 231–224.
65. Shahinzadeh, H., Moradi, J., Gharehpetian, G. B., Nafisi, H., & Abedi, M. (2019). IoT architecture for smart grids. In *International conference on protection and automation of power system (IPAPS)* (pp. 22–30). IEEE.

66. Tushar, W., Yuen, C., Saha, T. K., Morstyn, T., Chapman, A. C., Alam, M. J. E., et al. (2021). Peer-to-peer energy systems for connected communities: A review of recent advances and emerging challenges. *Applied Energy, 282*, 116131.
67. Baig, M. J. A., Iqbal, M. T., Jamil, M., & Khan, J. (2020). IoT and blockchain based peer to peer energy trading pilot platform. In *11th IEEE annual information technology, electronics and mobile communication conference (IEMCON)* (pp. 0402–0406). IEEE.
68. Yang, Q., & Wang, H. (2021). Privacy-preserving transactive energy management for IoT-aided smart homes via blockchain. *IEEE Internet of Things Journal, 8*(14), 11463–11475.
69. Moghaddam, M. H. Y., & Leon-Garcia, A. (2018). A fog-based internet of energy architecture for transactive energy management systems. *IEEE Internet of Things Journal, 5*(2), 1055–1069.
70. Mollah, M. B., Zhao, J., Niyato, D., Lam, K. Y., Zhang, X., Ghias, A. M., et al. (2020). Blockchain for future smart grid: A comprehensive survey. *IEEE Internet of Things Journal, 8*(1), 18–43.
71. Gai, K., Wu, Y., Zhu, L., Qiu, M., & Shen, M. (2019). Privacy-preserving energy trading using consortium blockchain in smart grid. *IEEE Transactions on Industrial Informatics, 15*(6), 3548–3558.
72. Yang, Q., & Wang, H. (2020). Blockchain-empowered socially optimal transactive energy system: Framework and implementation. *IEEE Transactions on Industrial Informatics, 17*(5), 3122–3132.
73. Siano, P., De Marco, G., Rolán, A., & Loia, V. (2019). A survey and evaluation of the potentials of distributed ledger technology for peer-to-peer transactive energy exchanges in local energy markets. *IEEE Systems Journal, 13*(3), 3454–3466.
74. Wu, Y., Wu, Y., Guerrero, J. M., & Vasquez, J. C. (2022). Decentralized transactive energy community in edge grid with positive buildings and interactive electric vehicles. *International Journal of Electrical Power & Energy Systems, 135*, 107510.
75. Curry, E., Hasan, S., Kouroupetroglou, C., Fabritius, W., ul Hassan, U., & Derguech, W. (2018). Internet of things enhanced user experience for smart water and energy management. *IEEE Internet Computing, 22*(1), 18–28.
76. Chen, X., Wang, X., Ni, W., & Collings, I. B. (2017). Two-way energy trading and online planning for fifth-generation communications with renewables. In *23rd Asia-Pacific conference on communications (APCC)* (pp. 1–6). IEEE.
77. Han, J., Liu, N., Huang, Y., & Zhou, Z. (2021). Collaborative optimization of distribution network and 5G mobile network with renewable energy sources in smart grid. *International Journal of Electrical Power & Energy Systems, 130*, 107027.
78. Hui, Y., et al. (2022). Secure and personalized edge computing services in 6G heterogeneous vehicular networks. *IEEE Internet of Things Journal, 9*(8), 5920–5931.
79. Kumari, A., Gupta, R., & Tanwar, S. (2021). PRS-P2P: A prosumer recommender system for secure P2P energy trading using Q-learning towards 6G. In *2021 IEEE international conference on communications workshops (ICC Workshops), Montreal, QC, Canada* (pp. 1–6). IEEE.
80. Ghoddusi, H., Creamer, G. G., & Rafizadeh, N. (2019). Machine learning in energy economics and finance: A review. *Energy Economics, 81*, 709–727.
81. Kell, A. J. M., McGough, S., & Forshaw, M. (2017). Machine learning applications for electricity market agent-based models: A systematic literature review. In *Conference17*, Washington, DC, USA.
82. PwC. (2018). *Sizing the prize. pwc's global artificial intelligence study: Exploiting the AI revolution*. Accessed: 26 Apr 2020. [Online]. Available from: https://pwc.com/gx/en/issues/data-and-analytics/publications/artificial-intelligence-study.html
83. Forootan, M. M., Larki, I., Zahedi, R., & Ahmadi, A. (2022). Machine learning and deep learning in energy systems: A review. *Sustainability, 14*(8), 4832.

84. Mohajeri, N., Assouline, D., Guiboud, B., Bill, A., Gudmundsson, A., & Scartezzini, J. L. (2018). A city-scale roof shape classification using machine learning for solar energy applications. *Renewable Energy, 121*, 81–93.
85. Dery, L. M., Nachman, B., Rubbo, F., & Schwartzman, A. (2017). Weakly supervised classification in high energy physics. *Journal of High Energy Physics, 2017*, 145.
86. Catalina, T., Iordache, V., & Caracaleanu, B. (2013). Multiple regression model for fast prediction of the heating energy demand. *Energy and Buildings, 57*, 302–312.
87. Haider, S. A., Sajid, M., & Iqbal, S. (2021). Forecasting hydrogen production potential in Islamabad from solar energy using water electrolysis. *International Journal of Hydrogen Energy, 46*, 1671–1681.
88. Cortes, C., & Vapnik, V. (1995). Support-vector networks. *Machine Learning, 20*, 273–297.
89. Seal, H. L. (1967). "Studies in the history of probability and statistics". XV The historical development of the Gauss linear model. *Biometrika, 54*, 1–24.
90. Hill, T., Marquez, L., Connor, M. O., & Remus, W. (1994). Artificial neural network models for forecasting and decision making. *International Journal of Forecasting, 10*, 5–15.
91. Alkhayat, G., & Mehmood, R. (2021). A review and taxonomy of wind and solar energy forecasting methods based on deep learning. *Energy AI, 4*, 100060.
92. Tschiatschek, S., Paul, K., & Pernkopf, F. (2014). Integer Bayesian network classifiers. *Lecture Notes in Computer Science, 8726 LNAI, PART 3*, 209–224.
93. Hartigan, J. A., & Wong, M. A. (1979). Algorithm AS 136: A K-means clustering algorithm. *Journal of the Royal Statistical Society C, 28*(1), 100–108.
94. Pinto, T., & Falcão-Reis, F. (2019). Strategic participation in competitive electricity markets: Internal versus sectorial data analysis. *International Journal of Electrical Power and Energy Systems, 108*, 432–444.
95. Fraunholz, C., Kraft, E., Keles, D., & Fichtner, W. (2021). Advanced price forecasting in agent-based electricity market simulation. *Applied Energy, 290*, 116688.
96. Pinto, T., Sousa, T. M., Praça, I., Vale, Z., & Morais, H. (2016). Support vector machines for decision support in electricity markets' strategic bidding. *Neurocomputing, 172*, 438–445.
97. Imran, K., Zhang, J., Pal, A., Khattak, A., Ullah, K., & Baig, S. M. (2020). Bilateral negotiations for electricity market by adaptive agent-tracking strategy. *Electric Power Systems Research, 186*, 1–12.
98. Gomes, I. L. R., Melicio, R., & Mendes, V. M. F. (2021). A novel microgrid support management system based on stochastic mixed-integer linear programming. *Energy, 223*, 120030.
99. Park, L., Jeong, S., Kim, J., & Cho, S. (2019). Joint geometric unsupervised learning and truthful auction for local energy market. *IEEE Transactions on Industrial Electronics, 66*(2), 1499–1508.
100. Sutton, R. S., & Barto, A. G. (2018). *Reinforcement learning: An introduction*. MIT Press.
101. Moerland, T. M., Broekens, J., & Jonker, C. M. (2020). Model-based reinforcement learning: A survey. arXiv preprint arXiv:2006.16712.
102. Wicaksono, A., & Prihatmanto, A. S. (2015). Optimal control system design for electric vehicle. In *2015 4th International conference on interactive digital media (ICIDM), Bandung, Indonesia*. IEEE.
103. Tucker, N., Moradipari, A., & Alizadeh, M. (2020). Constrained Thompson sampling for real-time electricity pricing with grid reliability constraints. *IEEE Transactions on Smart Grid, 11*(6), 4971–4983.
104. Li, Y., Guo, S., Zhu, L., Mukai, T., & Gan, Z. (2019). Enhanced probabilistic inference algorithm using probabilistic neural networks for learning control. *IEEE Access, 7*, 184457–184467.
105. Chen, X., Qu, G., Tang, Y., Low, S., & Li, N. (2021). Reinforcement learning for decision-making and control in power systems: Tutorial, review, and vision. arXiv:2102.01168.
106. Grondman, I., Busoniu, L., Lopes, G. A. D., & Babuska, R. (2012). A survey of actor-critic reinforcement learning: Standard and natural policy gradients. *IEEE Transactions on Systems, Man, and Cybernetics, Part C (Applications and Reviews), 42*(6), 1291–1307.

107. Yang, K.-E., et al. (2021). Trust-region method with deep reinforcement learning in analog design space exploration. In *2021 58th ACM/IEEE design automation conference (DAC)* (pp. 1225–1230). IEEE.
108. Tedrake, R., Zhang, T. W., & Seung, H. S. (2004). Stochastic policy gradient reinforcement learning on a simple 3D biped. In *2004 IEEE/RSJ international conference on intelligent robots and systems (IROS) (IEEE Cat. No.04CH37566)* (Vol. 3, pp. 2849–2854). IEEE.
109. Wu, Y., Tan, H., Peng, J., Zhang, H., & He, H. (2019). Deep reinforcement learning of energy management with continuous control strategy and traffic information for a series-parallel plug-in hybrid electric bus. *Applied Energy, 247*, 454–466.
110. Wang, D., Liu, B., Jia, H., Zhang, Z., Chen, J., & Huang, D. (2020). Peer-to-peer electricity transaction decision of user-side smart energy system based on SARSA reinforcement learning method. *CSEE Journal of Power and Energy Systems, 8*(3), 826–837.
111. Jang, B., Kim, M., Harerimana, G., & Kim, J. W. (2019). Q-learning algorithms: A comprehensive classification and applications. *IEEE Access, 7*, 133653–133667.
112. Vankadari, M. B., Das, K., Shinde, C., & Kumar, S. (2018). A reinforcement learning approach for autonomous control and landing of a quadrotor. In *2018 International conference on unmanned aircraft systems (ICUAS), Dallas, TX, USA* (pp. 676–683). IEEE.
113. Mbuwir, B. V., Kaffash, M., & Deconinck, G. (2018). Battery scheduling in a residential multi-carrier energy system using reinforcement learning. In *2018 IEEE international conference on communications, control, and computing technologies for smart grids (SmartGridComm), Aalborg, Denmark* (pp. 1–6). IEEE.
114. Cheung, W. C., Simchi-Levi, D., & Zhu, R. (2020). Reinforcement learning for non-stationary markov decision processes: The blessing of (more) optimism. In *Proc. Int. Conf. Mach. Learn.* PMLR, pp. 1843–1854.
115. Jamil, F., Iqbal, N., Imran, Ahmad, S., & Kim, D. (2021). Peer-to-peer energy trading mechanism based on blockchain and machine learning for sustainable electrical power supply in smart grid. *IEEE Access, 9*, 39193–39217.
116. Kim, J. G., & Lee, B. (2020). Automatic P2P energy trading model based on reinforcement learning using long short-term delayed reward. *Energies, 13*(20), 5359.
117. Guo, C., Wang, X., Zheng, Y., & Zhang, F. (2021). Optimal energy management of multi-microgrids connected to distribution system based on deep reinforcement learning. *International Journal of Electrical Power & Energy Systems, 131*, 107048.
118. Zang, H., & Kim, J. (2021). Reinforcement learning based peer-to-peer energy trade management using community energy storage in local energy market. *Energies, 14*, 4131.
119. Taghizadeh, A., Montazeri, M., & Kebriaei, H. (2022). Deep reinforcement learning-aided bidding strategies for transactive energy market. *IEEE Systems Journal, 16*, 4445.
120. Chen, T., & Su, W. (2018). Local energy trading behavior modeling with deep reinforcement learning. *IEEE Access, 6*, 62806–62814.
121. Zhang, S., May, D., Gül, M., & Musilek, P. (2022). Reinforcement learning-driven local transactive energy market for distributed energy resources. *Energy and AI, 8*, 100150.
122. Yan, L., Chen, X., Chen, Y., & Wen, J. (2022). A hierarchical deep reinforcement learning-based community energy trading scheme for a neighborhood of smart households. *IEEE Transactions on Smart Grid, 13*, 4747.
123. Gao, G., Wen, Y., & Tao, D. (2022). Distributed energy trading and scheduling among microgrids via multiagent reinforcement learning. In *IEEE transactions on neural networks and learning systems*. IEEE.
124. Ye, Y., Papadaskalopoulos, D., Yuan, Q., Tang, Y., & Strbac, G. (2022). Multi-agent deep reinforcement learning for coordinated energy trading and flexibility services provision in local electricity markets. *IEEE Transactions on Smart Grid, 14*, 1541.
125. Zhao, Z., Feng, C., & Liu, A. L. (2022). Comparisons of auction designs through multi-agent learning in peer-to-peer energy trading. *IEEE Transactions on Smart Grid, 14*, 593.
126. Wang, J., Li, L., & Zhang, J. (2022). Deep reinforcement learning for energy trading and load scheduling in residential peer-to-peer energy trading market. *SSRN*.

# Chapter 4
# Blockchain-Based Transaction Platform for Peer-to-Peer Energy Trading

**Mehdi Zeraati, Farkhondeh Jabari, and Saeed Salarkheili**

## 4.1 Introduction

The global energy system is experiencing a huge transition from fossil fuel conditions to renewable energies. A significant part of greenhouse gases is produced by the energy sector, and it is necessary to take immediate action for climate change and environmental issues. The development of digitalization in the energy system plays an essential role in deploying innovations and new advanced technologies in the planning and operation areas, and in legislative and business models.

To date, the most widely used structure of the power grid is the top-down method. After power generation in power plants, it is distributed among consumers by transformers and transmission lines, so that the power flow is unidirectional from production to consumption. In the classic model of power system, a producer, through transmission lines, feeds a distribution network to which consumers are connected. It should be noted that a typical consumer gets all the required energy from the main grid. However, there is another group of consumers who produce energy mainly from renewable sources and are known as prosumers. At first, prosumers often provide only a part of their energy needs, but now they may sell their excess production to the grid.

Recent developments in power system technologies, such as distributed energy resources (DER) and two-way communication, have led to the restructuring and

M. Zeraati · F. Jabari (✉)
Power Systems Operation and Planning Research Department, Niroo Research Institute (NRI), Tehran, Iran
e-mail: mzeraati@nri.ac.ir; fjabbari@nri.ac.ir

S. Salarkheili
Faculty of Electrical Engineering, Shahid Beheshti University, Tehran, Iran
e-mail: s_salarkheili@sbu.ac.ir

 
V. Vahidinasab, B. Mohammadi-Ivatloo (eds.), *Demand-Side Peer-to-Peer Energy Trading*, Green Energy and Technology, https://doi.org/10.1007/978-3-031-35233-1_4

modernization of power systems [1]. One of the recent trends in the electricity market is the focus on energy self-sufficiency through small-scale power plants directly installed at the consumers' locations [2]. These installations often include rooftop solar panels and small wind turbines. New technologies and cost reductions in sustainable power generation and energy storage units have made power generation attractive to consumers [3]. As a result, the number of actors participating in small-scale energy generation is increasing [4]. This increases the decentralized generation, which is in contrast to a small number of centralized power producers with enormous energy production to cover all energy demand [3].

On the other hand, the increase in DERs has brought many social and economic improvements in different parts of the world. In particular, autonomous microgrids (MGs) play a vital role in such situations. MGs offer a worthy alternative to high-capacity power transmission systems and, instead of centralized power generation and then transmitting it over long vulnerable paths, they use distributed generation near the point of consumption.

In today's microgrids, various elements such as DERs, distributed generators (DG), energy storage systems, and demand response programs are used for the local supply of electrical loads, which necessitates data exchange between different parts of the MG to enable technical and economic decision-making for network operation and control. Modern power distribution systems are expected to intelligently implement these technologies and provide clean and cheap energy with high reliability to their customers. The availability of telecommunication infrastructure for information exchange helps distribution system operators (DSO) to perform the following actions [5]:

- Implementation of real-time monitoring and control
- Better management of outages and emergencies
- Facilitating conditions for increasing the penetration of DERs
- Reducing the operation costs
- Improving efficiency, reliability, and stability
- Reduction of peak consumption
- Minimizing the reservation requirements

However, modern microgrids pose significant challenges for DSOs with the increasing penetration of DERs:

- Vulnerability of telecommunication systems against malicious attacks
- Various security and privacy threats
- Scalability of control and management methods
- Resiliency against different types of disturbances
- Fault tolerance and adaptability

Current mechanisms and rules of electricity markets do not consider the significant number of self-sufficient users and do not offer any incentive for their development and expansion. In most cases, the DGs must inject power into the public grid for a specific price set by the local power distributor (LPD) or be consumed by themselves [6]. One of the goals of market deregulation is to allow customers to choose their energy supplier freely. These rules provide attractive incentives for sharing energy between different entities participating in the electricity market [7].

The smart grid (SG) is a concept in a power system where all parties collaborate to achieve a common goal of providing a sustainable, economic, and secure power supply [8]. SG connects heterogeneous components that have different functions and requirements. The increasing utilization of electric vehicles (EVs), the global trend towards low-carbon energy solutions, and the desire for RESs have made the SG's control and management methods more difficult and complex. In addition, the amount of energy exchanged and the number of network participants is increasing rapidly. All these changes necessitate cyber security and more excellent network stability.

In complex SGs, many transactions are performed between different entities simultaneously. One of the main concerns is transaction validation between various entities involved in a particular SG application. Recently, peer-to-peer (P2P) trading has attracted much attention as an energy management paradigm that can activate customers in the electricity market. Local P2P markets are usually established in a local community where all the community members are fed with the same voltage level, downstream of a transformer. This idea may lead to possibly lower costs for consumers and more incredible benefits for prosumers [9].

There are studies to add features such as unified payments interface (UPI) and distributed ledger technologies (DLT) to enable real-time energy transactions between prosumers [10]. This is enabled by the advances in SGs, MGs, modular DERs, and information and communication technologies (ICT) [4]. With the help of these developments, along with the proposed changes in structure and operation, the power systems are expected to move towards smart microgrids, where the real-time bidirectional flow of power, money, and information is practical. These interconnected smart microgrids also provide higher energy reliability and security during power outages, shortages, and cyberattacks because these networks, in cooperation with each other, can act as a reservation and provide recovery services [11].

As mentioned, interconnected smart microgrids need the advanced data management and telecommunication systems to operate effectively. By exploiting decentralized DGs, their databases and management systems must also be distributed. These decentralized systems have more flexibility and speed than centralized types [11]. DLT is a decentralized digital database widely applied in distributed environments where information is shared among many users simultaneously [12]. Blockchain is one of the most well-known types of DLT, which has attracted a lot of attention in many industries, such as finance, supply chain, energy, and the Internet of Things (IoT).

In this chapter, the applications of blockchain technologies in power distribution systems are introduced. The most important parts of a blockchain in these systems are discussed, and the challenges of their practical implementation are examined. Section 4.2 is dedicated to presenting the idea of peer-to-peer (P2P) energy trading. In Sect. 4.3, blockchain technologies are introduced. In Sect. 4.4, security and privacy of smart energy networks are discussed. In Sect. 4.5, blockchain weaknesses and vulnerabilities are presented. Finally, this chapter concludes in Sect. 4.6.

## 4.2 Peer-to-Peer (P2P) Energy Trading

There is great potential for activating small players in decentralized energy markets. Energy consumers can sell energy to the grid as prosumers and exchange energy with each other, as shown in Fig. 4.1. For this purpose, creative business models are needed to minimize the total costs and simultaneously maximize the customers' benefits. This model stands in contrast to the conventional vertical structure and creates a horizontal structure for sharing, transferring, exchanging, and trading energy between customers [5].

P2P energy trading is an innovative energy transfer model that allows interconnected MGs to exchange energy flexibly. In this case, MGs are considered a controlled load from the DSO viewpoint, which can be used to overcome the variable output of DER units. The introduction of P2P energy trading in modern power distribution systems enables the decentralized and active operation of local MGs.

Undoubtedly, P2P energy trading is considered one of the main mechanisms of future energy management. For P2P networks, transactive management enables decentralized control of participating parties in the energy network compared to conventional energy networks, where they are centrally managed [13]. Communication between participants in a central mechanism must be realized by main servers, and subsequently, the number of equipment required increases with the number of participants [9]. Therefore, this model is not easily scalable to implement, considering the exponential growth of customers. High implementation costs make this method economically impossible [13]. Instead, P2P transactions can provide a cheaper mechanism than the traditional methods based on the central information system.

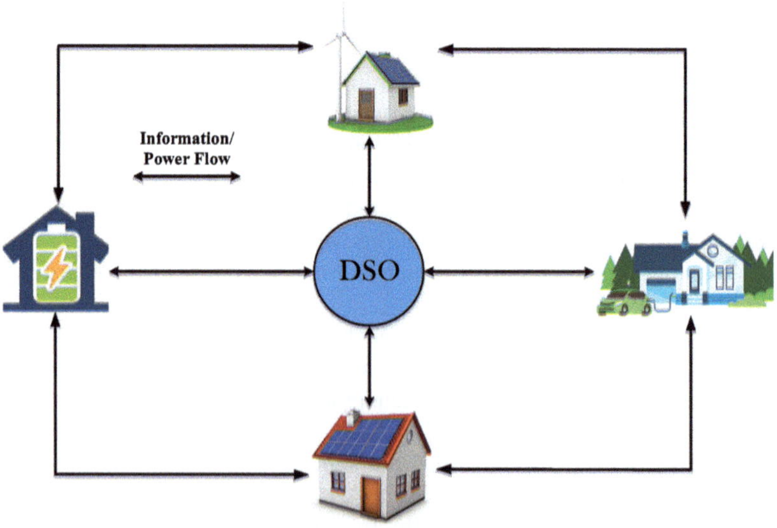


**Fig. 4.1** Peer-to-Peer energy trading between prosumers in smart grid

However, energy trading in a P2P network is challenging because prosumers need to generate their energy without the planning and supervision of a central authority. This may make P2P markets an untrustworthy system. In addition, to enable the P2P energy market, various smart grid applications, including real-time pricing, advanced metering, and demand-side management, are combined into a single function [14]. On the other hand, the P2P energy market collects information about buy and sell orders in a period and processes this data in a decentralized platform. Therefore, the applied communication technologies should be able to perform these tasks in the P2P energy market environment.

In P2P energy trading, MGs determine their trading patterns based on market prices and other criteria that reflect their specific goals. This type of business offers valuable solutions to deal with technical and economic challenges in power distribution systems. In addition, it encourages MGs to maximize their local energy production to provide affordable energy prices to their customers. Blockchain enables decentralized P2P transactions, which is entirely in line with the idea of conducting energy transactions in the local energy market without the need for a central authority such as the DSO in conventional power distribution networks [15], which is discussed in the following subsection.

Some pros and cons of P2P trading platform can be stated as follows:

- *Advantages*
  - Exchanging electricity between consumers and producers without the need for a third person or entity.
  - Prosumers can sell surplus electricity to peers at higher market prices.
  - Higher resiliency of energy systems using more accessible distributed generation units.
  - Balancing and congestion management.
  - Providing ancillary services to main power network.
  - Application of demand-side management programs.
  - Enabling standalone operation of microgrids and nanogrids.
  - Reducing on-peak electricity demand and transmission losses.
  - Application of blockchain technology in decentralized communication infrastructure.
  - Smart contracts for automated business models.
  - Robust and risk-averse mechanism against uncertainties.
- *Disadvantages*
  - Cyber threats of smart contracts.
  - Higher investment cost of blockchain technology.
  - Participants with low knowledge about dynamic energy market may submit unreasonable bids.
  - Huge volume of data transferring and processing at each second.

## 4.3 Blockchain Technology

Blockchain is a growing technology that is an alternative to conventional economic trading systems and has the potential to create more homogeneous business models across various industry sectors. Blockchain is a peer-to-peer distributed ledger that enables the trading or sale of assets with complete tracking of the transaction without the need for a central authority. In addition, it facilitates the sharing and integration of information across the network in a distributed and secure framework. Initially, blockchain was designed to perform peer-to-peer financial transactions in a completely decentralized way. This technology includes LTDs that store transaction information with high security without the presence of a centralized entity. Furthermore, they allow the real-time execution of smart contracts and ensure fast settlement of payments [16]. Due to these advantages, the utilization of blockchain in the energy sector is increasing rapidly (Fig. 4.2).

In this method, each transaction is kept in a block on the network. A block, similar to a chain structure, stores the hash value of the previous block (Fig. 4.3). This structure creates more consistency. Each transaction on the blockchain can be represented using cryptographically signed blocks, and transactions are validated by network users [17]. Various consensus algorithms are used by blockchain to verify transactions. Consensus algorithms are agreements implemented between a group of people to confirm transactions. The decision is made by a majority vote at the end of the validation process [18]. Smart contracts are also essential components in many blockchain platforms. A smart contract is a set of rules that apply to a

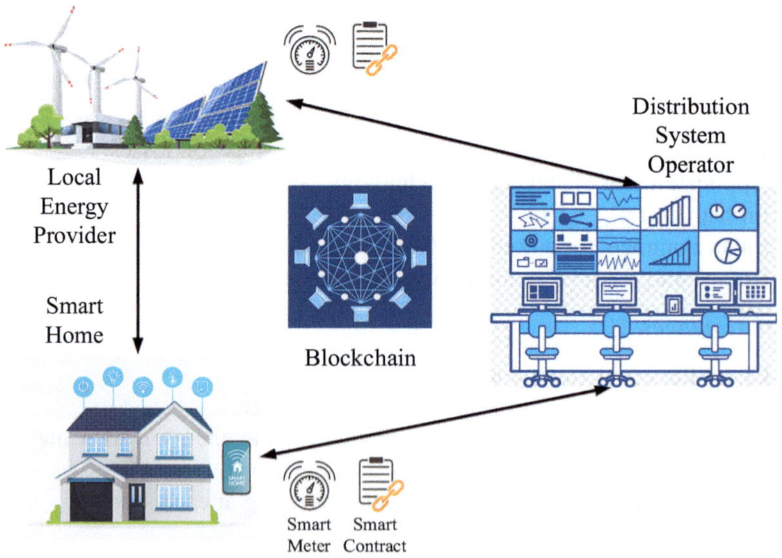


**Fig. 4.2** Blockchain in smart grids

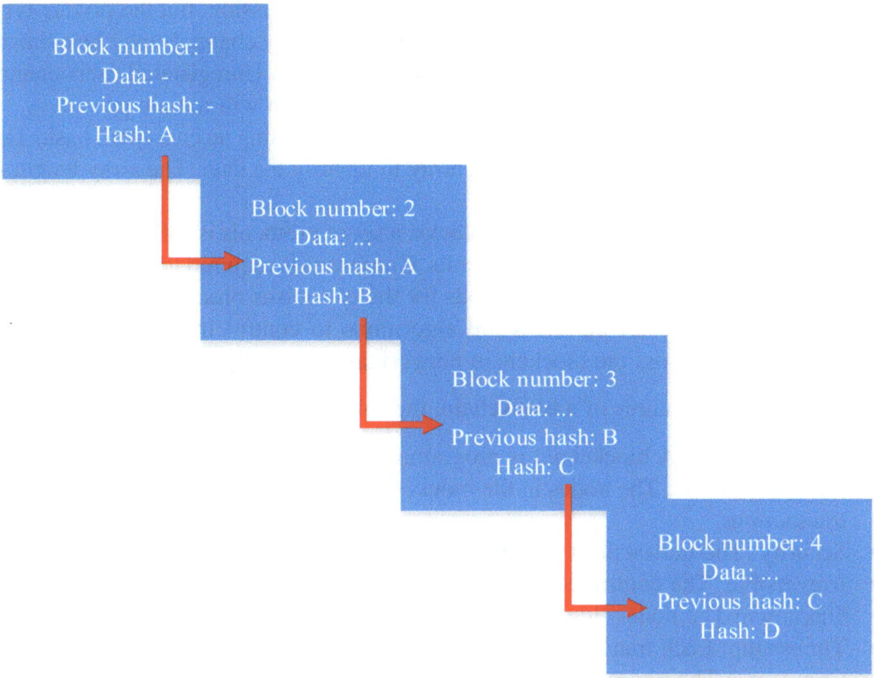


**Fig. 4.3** Blockchain structure

blockchain. As a software representative of users, the smart contract automatically performs certain obligations and tasks in response to existing conditions [19].

In other words, in order to control and manage decentralized P2P approaches in smart grids, there is a need to develop decentralized management methods. Blockchain technology is considered a suitable solution to deal with these challenges in modern power distribution systems. The introduction of blockchain provides a reliable mechanism for direct interaction between participants and securely enables P2P transactions. In addition, blockchain technology offers faster and more transparent system operations. As a result, blockchains can be used to improve security and efficiency in power distribution systems.

The most important components of a blockchain are:

*Block*: In a blockchain, the blocks are represented by pointers and linked list data structures and are stored in a logical order using a linked list. A block is a set of data containing transaction information, such as timestamps and links to previous blocks. These data are produced using a secure hashing technique. The pointers dedicate the location of the next block.

*Public and private keys*: Blockchain is a network of interconnected and secured blocks constantly growing using cryptographic processes [20]. In order to verify transaction authorization, blockchain utilizes an asymmetric key technique. The transactions in the block are encrypted with a private key. The public type is transparent and can be seen by everyone without permission, but the private type is only accessible to authorized users. Therefore, many channels may be created, and a certain number of users could be linked to them. Unregistered users cannot access the data. Additionally, confidential information will remain private.

*Hash function*: Each block relates to the previous block using an encrypted hash. The hash function creates a fixed-length array to identify the data. The array length is independent of the size of the data.

*Consensus Process*: To validate new blocks, a set of protocols is followed to reach a consensus from all network participants. Consensus is required to decide on block validation. There are several methods for the consensus process.

*Smart Contracts*: Smart contracts are algorithms to control transactions between nodes spread across the blockchain network automatically.

Among other features of a blockchain, the following can be mentioned:

- Programmable: A blockchain is programmable.
- Decentralization: The nodes in the blockchain network have access to each other transactions.
- Security: The transactions are secured using hash functions.
- Irreversible: The confirmed transactions cannot be reversed.
- Unknown: The nodes' identity is anonymous.
- Timestamp: Each transaction has a timestamp.
- Consensus: The nodes must agree for a transaction to be confirmed.

As mentioned, blockchain, as a trusted decentralized ledger technology, provides excellent potential for establishing communication among distributed nodes in the network and provides a cooperative network for P2P energy/data transactions between untrusted parties. This technology breaks down the limitations of centralized supply chains and enables a wide range of flexible energy transactions between multiple sectors. In addition to P2P trading, blockchain improves the reliability, resiliency, and cyber security of the network. In the following, the layered architecture of the blockchain is introduced in the format of bottom-up design:

*Data Management Layer*: This layer is responsible for data structure, encryption, and storage. The data management layer often consists of data blocks, blockchains, and cryptographic algorithms.

*Networked Layer*: This layer is designed for node management, P2P broadcasting, privacy control, and security. It usually consists of P2P network protocols, policy configuration, security, and data privacy.

*Consensus layer*: This layer is known as the core of the blockchain network. The consensus algorithm is responsible to ensure that the nodes reach a specific agreement. Additionally, blocks validation, synchronization of all nodes, and ensuring that only one chain is maintained are other functionalities of this layer.

*Incentives layer*: Incentive models and reward mechanisms are needed to promote the development and maintenance of the blockchain system. This layer is critical for the blockchain network to remain stable and attractive.

*Services Layer*: This layer handles business process automation, identity management, and distributed storage certification. It often consists of wallets, certificate authorization, digital identity determination, smart contracts, and channel management for permissioned blockchain networks.

*Application Layer*: The interface for users to interact with the blockchain network is realized in this layer. Moreover, invoking smart contracts, engaging in role-based activities, and accessing distributed ledgers are examples of other tasks of the application layer.

Various multi-layer architectures have been proposed in literature for the blockchain structure and the P2P energy market in smart grids. For example, Fig. 4.4 illustrates a four-layer structure in which the role of each layer is as follows:

*Infrastructure layer*: This layer consists of physical equipment, connections, meters, and control systems. Power flow directions and energy exchanges are physically defined in this layer. The recorded measurements are collected by smart meters and sent to the upper layers.

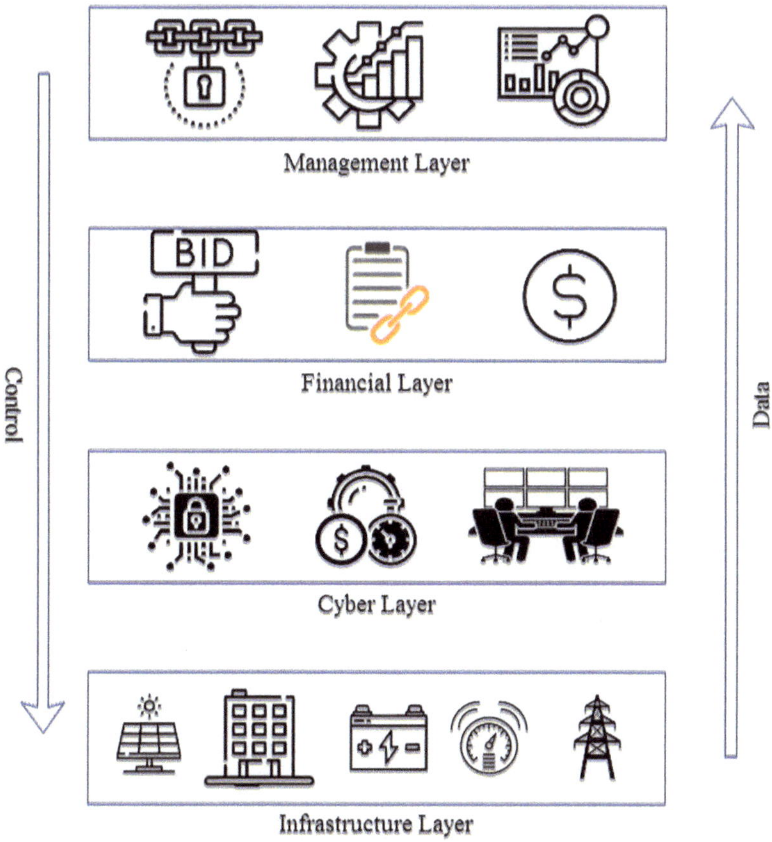


**Fig. 4.4** Sample architecture for blockchain structure and P2P energy trading in smart grids

*Cyber layer*: In this layer, it is ensured that the system is operated in a safe mode. Constraints such as generation and demand balance and voltage and current limits for physical equipment must be satisfied in all cases. This layer ensures the safe operation of system in normal conditions despite P2P exchanges. In addition, if a constraint violation occurs, necessary correction commands are sent to the infrastructure layer to fix it.

*Financial layer*: The auction mechanism including bids from buyers and offers from sellers is run in this layer. Therefore, market clearing and price calculation are also done in this layer based on the information received from the lower layers.

*Management layer*: This layer facilitates the implementation of blockchain mechanism and smart contracts. It is also responsible for processing data received from lower layers, making decisions, and issuing necessary control commands.

## 4.4 Security and Privacy of Smart Energy Networks

Data security and privacy of smart grids are two main factors for their continuous, reliable, and stable operations. Recently, electricity networks have faced various cyber-physical threats such as denial of service (DoS) attacks, false data injection (FDI) risks, and time-delay intrusions. Some of recent cyberattacks to smart systems are presented in Table 4.1 [21].

Recently, energy systems are being more safe, secure, and energy efficient using blockchain, which integrates some information and communications technologies (ICT) with monitoring, protection, and control centers. There are different smart devices in modern energy grids to optimally manage generation, transportation, distribution, and consumption of various energy forms such as cooling, heating, electricity, fresh water procurement, etc. Data security and privacy are required for reliable, stable, and safe operation of these grids. In the centralized management of the smart grids, attackers can maximize the intrusion impacts on monitoring, protection, and control units by any failure in the control center. Hence, it is a crucial challenge for system operator to make appropriate decisions in efficient management of centralized energy networks. In recent years, the number of prosumers have been increased, which leads to a significant increase in cost and computational burden of centralized management systems. Therefore, implementation of demand-side management programs has faced some problems. For example, some customers are interested in purchasing electrical power from local prosumers directly, which causes management issues. Penetration of renewable energy resources based on distributed generations demonstrates that decentralized management of energy grids is unavoidable. In addition, unreasonable monopoly of energy market prices may be unexpected in centralized transactions. As a result, efficient and reliable energy trading should be realized in smart grids [23].

In the centralized energy systems, data security and privacy concerns are growing. Malicious third-party attacks and intrusions and data privacy leakage may lead to severe security damage and higher irreparable financial losses. Hence, the optimal

**Table 4.1** Recent cyberattacks to energy systems of different countries [22]

| Attack source | Attack target | Impacts on power system data quality | System weaknesses | Year |
|---|---|---|---|---|
| Vulnerability of firewall network | Web servers of California's ISO | Data accuracy and consistency have been affected. | Security consideration under system maintenance was poor. | 2001 |
| Stuxnet worm | Programmable logic controllers in supervisory control and data acquisition (SCADA) | Data accuracy, consistency, and reasonableness have been affected. | Vulnerabilities of programmable logic controllers was high. | 2010 |
| BlackEnergy malware | In Ukraine, multiple companies working with ICS-CERT have identified this malware on internet-connected human-machine interfaces (HMIs). | Data origin, accuracy, and consistency have been affected in HMI of control center. | Attack target was general electric's HMI. | 2011 |
| Watering Hole Trojan | Industrial control system of SCADA | Data confidentiality, integrity, and availability were targeted. | Infecting user's computer and gaining access to network. | 2014 |
| Laziok Trojan | Energy companies were targeted. | Data origin and confidentiality were affected. | Vulnerable devices in system | 2015 |
| WannaCry ransomware attack | A worldwide cyberattack which targeted computers running Microsoft Windows operating system by encrypting data and demanding ransom payments in Bitcoin cryptocurrency. | Data availability was affected. | It is considered a network worm because it also includes a transport mechanism to automatically spread itself. This transport code scans for vulnerable systems, then uses EternalBlue exploit to gain access, and DoublePulsar tool to install and execute a copy of itself | 2017 |

management of the centralized energy grids with a large number of monitoring, control, and protection devices is challenging. Meanwhile, there are several limitations in metering a huge number of data, recording in a database, transferring via communication systems, and gathering them by a central server. In recent years, microgrids have attracted world attention because of their potential in local energy generation, distribution, and consumption with lower power losses and less need of long transmission lines from generation to consumption levels. However, there are still some unsolved issues in data security and privacy of energy trading platforms. Electrification of transportation has faced microgrids with other cyber security issues due to vulnerability of electric vehicles against external attacks, especially in considering vehicle-to-grid (V2G) operating modes of pure electric or hybrid vehicles and road conditions [24, 25].

Transformation of centralized energy management approaches to distributed or decentralized one is essential, which can be done using blockchain. Blockchain technology makes it possible to balance load generation, implement demand-side management strategies, and dispatch loads between various distributed energy resources such as renewable energies or fossil fuel based ones. Due to higher flexibility of blockchain-enabled smart grids, end-users or consumers can participate in demand response programs as well as energy generation and storage through smart contracts, and they can be called self-reliant customers or prosumers. A smart grid can be changed into a decentralized energy trading platform with multiple management points [26].

Blockchain is classified into three main categories: private, public, and consortium. Everyone can participate in public blockchain. In a consortium blockchain, only coalition members are invited to participate in trading process. A private blockchain is a distributed ledger that securely operates using cryptographic concepts based on an organization's or a person's requirements. It is a practical solution for energy trading in modern electricity networks with capability of analyzing data/cash/energy flows. One of reasons for the monopoly operation of energy trading in conventional systems is their centralized structure. In the blockchain-enabled decentralized energy management, prosumers can participate in peer-to-peer energy trading. Immutability is related to cryptography and hashing process of blockchain. A hash function receives data as inputs and generates a digital signature as checksum or output for ensuring data correctness. All transactions which are verified in blockchain will be timestamped and recorded in a secure information block linking to previous block's output. Therefore, this chain of blocks will be unbreakable and data deletion or manipulation probability converges to zero. Transparency of blockchain refers to accessibility, informativeness, understandability, and auditability of P2P transactions as shown in Fig. 4.5.

According to accessibility property of blockchain-enabled transactions, all information is encrypted and "available" while these data are stored in multiple nodes, aiming to reduce information deletion or manipulation. Portability denotes accessibility via multiple sources and consistency/adaptability with different programming languages such as C++, JavaScipt, Solidity, Python, PHP, C#, Ruby, Rust, SQL, Erlang, and Rholang. Standardization and use of protocols add uniformity to blockchain-enabled P2P transactions. Moreover, transaction validation is carried out in multiple nodes, which denotes to operability of blockchain-based P2P energy trades. Simplicity and user-friendliness are provided by availability of more detailed information in each block of chain. Verification of information of each block with those available in other network nodes ensures their accuracy and correctness. All information is stored in blocks making it possible to verify and trace them during transactions [27].

In the decentralized architecture of the smart grids, P2P energy trading is facilitated between consumers, distributed generation units, storages, demand response participants, and prosumers. All transactions can be recorded and updated in all authorized nodes while satisfying data integrity. Smart contracts make it possible to execute P2P energy trades automatically in smart grids. Smart grids compose of physical equipment, communication and control infrastructure, application

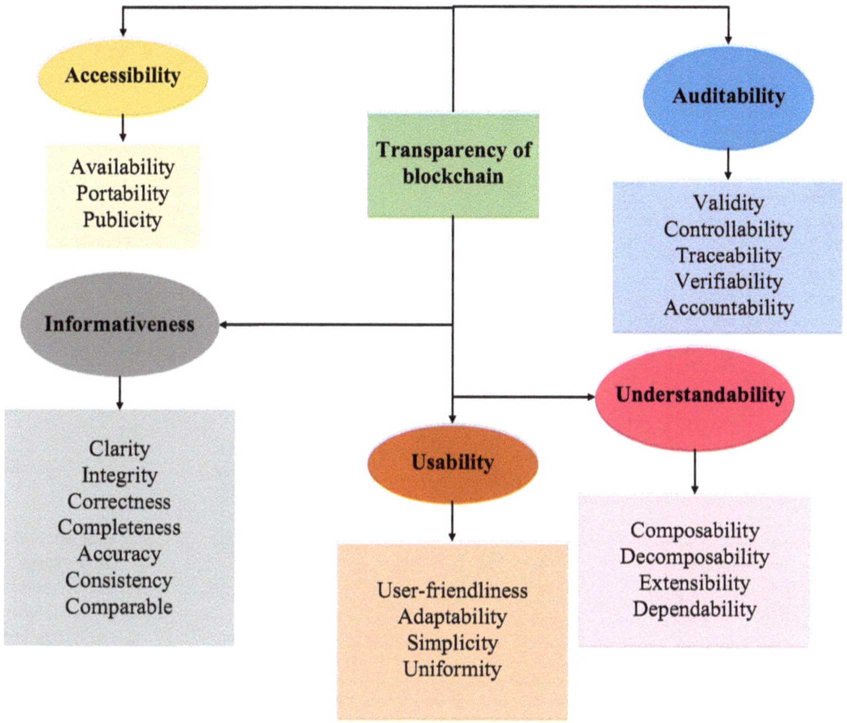


**Fig. 4.5** Transparency criteria of blockchain technology

functions, data gathering, and analytic layers, as shown in Fig. 4.6. The first layer of the smart grid consists of measuring units and sensors for real-time monitoring, management, and control. Some of these smart metering devices are micro-phasor measurement units (μPMUs), remote terminal units (RTUs), smart power quality meters, advanced metering infrastructure (AMI), current and voltage transformers (CVTs), and wireless sensor network (WSN) which communicate data. Communication systems and protocols provide a reliable and secure manner for real-time data gathering, recording, transferring, processing, and analysis, which can be used by monitoring and control modules. All transactions between various components of smart grids, such as small and large power generation units, storage technologies, demand response participant, pure or hybrid electric vehicles, etc., can be carried out in application layer. Moreover, AMI facilitates data gathering and transferring between multiple nodes. Blockchain is used for P2P energy trading in smart grids to ensure that all participants are trusted and transactions are secure. Moreover, transparency of all smart contracts is guaranteed [28, 29].

Application of internet-of-things (IoT) in energy grids added cyber-physical security issue to smart grids, which is classified into five categories as Fig. 4.7. Physical attacks consist of sensors and smart meters tampering aiming to achieve important or vital information, jamming nodes in WSN, denying data

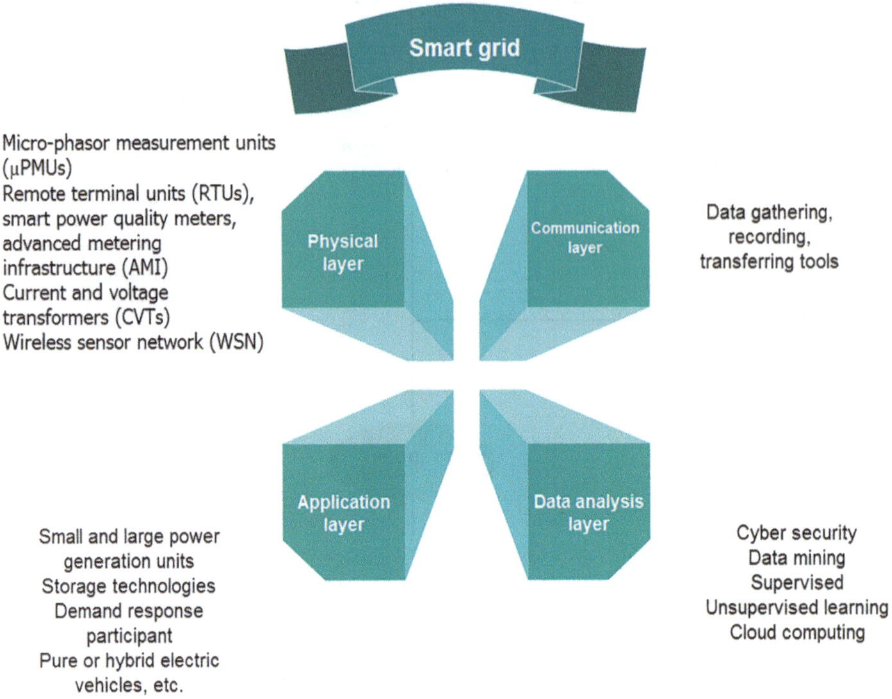


**Fig. 4.6** Schematic representation of smart grid layers

communication, and injecting malicious or false data. All of these attacks affect the distribution system state estimation accuracy. In the software targeted attacks, authenticity of user, viruses, and malicious programs are used for data theft, tampering, and DoS. The network attacks aim to access confidential data, and unauthorized access to sensitive information. DoS intrusion, FDI attacks, and injecting malicious control and management signals are some of the cyberattacks to system controllers. Data confidentiality is another type of encryption attack. To exchange data in a secure and confident manner and ensure integrity, accuracy, availability, and consistency of data, blockchain is applied to smart grids [30].

In a decentralized energy grid, if a part the of system fails, others will normally be operated. Moreover, all transaction information is communicated between consumers and prosumers through a blockchain-enabled platform. When all transactions are confirmed, contracted power will be transmitted from sellers to consumers via the physical components of the smart grids. All transactions are recorded and authenticated by participants, stakeholders, and entities (peers) using decentralized consensus. Cryptographic solutions are considered during smart energy trading to ensure the ledger data security. As a result of transparency feature of blockchain-enabled transactions, data tampering, and information tracing will be difficult and easy, respectively. In Table 4.2, some applications are stated for blockchain technology in smart energy networks.

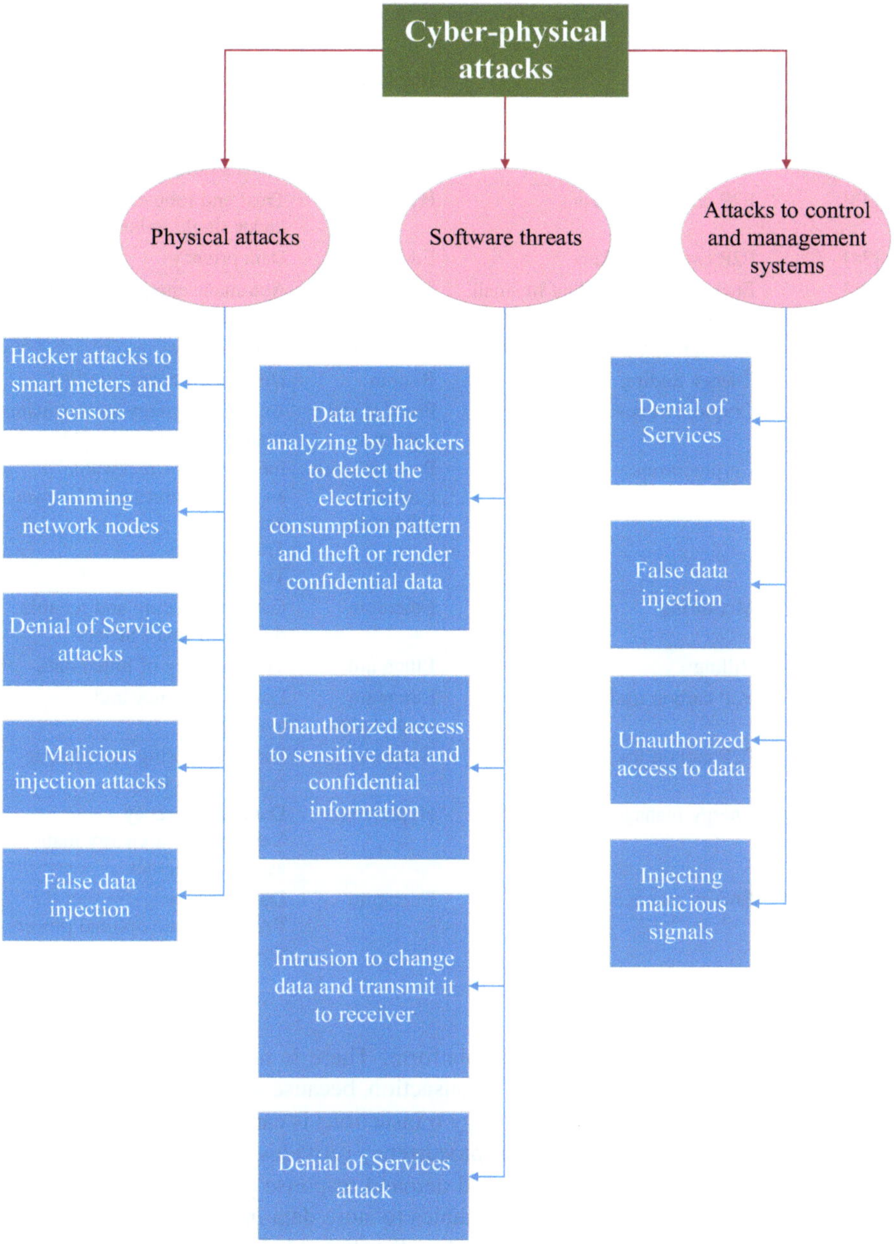


**Fig. 4.7** Cyber-physical attacks in smart grids

According to Table 4.2, P2P energy trading, energy markets in microgrid scale, cryptocurrency, metering and billing systems, carbon trading, energy management, IoT, and electric transportation are the main applications of blockchain technology. In a smart distribution system, multiple microgrids can securely trade energy based

**Table 4.2** Security and privacy issues in recent applications of blockchain-enabled smart energy transactions

| References | Application | Blockchain type | Objective(s) |
|---|---|---|---|
| [32] | P2P energy markets | Hyperledger | Data transparency<br>Transactions authentication |
| [33] | P2P energy markets | Hyperledger | Trust and robust operation under single failure |
| [34] | P2P energy markets | Ethereum | Data privacy |
| [35] | Energy market trading in small or microgrid communities | Ethereum | Automatic energy trading using smart contracts |
| [36] | Microgrid-scale energy trading | Hyperledger | Data privacy |
| [37] | Energy trading | Bitcoin | Distributed energy trading |
| [38] | Cryptocurrency | Proof of stake consensus | Automatic energy trading using smart contracts |
| [35, 39] | Cryptocurrency | Proof of work consensus<br>Ethereum | Incentives for renewable energy resources based power generation<br>Transparency<br>Trust |
| [40] | Measuring | Ethereum | Gathering secure and reliable data from smart meters |
| [41] | Billing | Ethereum | Transparency of billed data |
| [42] | P2P carbon trading | Ethereum<br>Hyperledger | Data consistency and availability |
| [43] | P2P carbon trading | Ethereum | Data monitoring for transparency and security |
| [44] | Energy management | Hyperledger | Data transparency<br>Verification of energy management schemes |
| [15] | Energy management | Ethereum | Data security<br>Run automatic optimal power flow algorithm |

on blockchain-enabled decentralized platform. There is no need for a centralized authority to run a cryptocurrency P2P transaction, because it is a blockchain oriented digital asset. Any payment in P2P energy transactions is carried out using Bitcoin or other cryptocurrencies. Moreover, cryptocurrency provides an incentive solution for prosumers, green energy producers, and demand response participants. In metering and billing systems, blockchain also enables to store data in a decentralized ledger. All measurements or billed data recorded in distributed ledger are authenticated and traced, which improves transparency and integrity of these systems. Blockchain-oriented database has tamper-proof feature, which makes it more reliable and secure for control and energy management actions. Penetration of electric drive vehicles has caused cyber-security challenges in transactions between electric vehicles' owners and charging stations. Meanwhile, blockchain enhances privacy and authentication of electric vehicles P2P energy transactions. According to consistency feature of blockchain-based energy transactions, a distributed ledger is simultaneously available for all nodes or participants. If there is a trade-off between consistency and availability properties, all read/write points will have access to updated data, reducing latency and probability of old information return. Ethereum and Bitcoin that use PoW consensus, have recently been used in energy transactions. For solving a computationally the hash process named "mining," annual electrical power consumption of Bitcoin varies between 60 and 150 TWh. Hence, high electricity requirement of Bitcoin transactions is one of their drawbacks. Each network node of blockchain-based decentralized energy trading platforms stores data in a distributed ledger. Therefore, a huge number of data should be stored for each node, which makes it expensive to synchronize data in real-time intervals [31].

## 4.5 Blockchain Weaknesses and Vulnerabilities

Although blockchain has several advantages in P2P energy trading platforms, some vulnerabilities and threats have been reported for Bitcoin-based transactions during 2011–2019, as shown in Fig. 4.8. Bitcoin client's vulnerabilities have mainly occurred due to digital signature threats, hash function vulnerability, user address vulnerability, mining process malware, and software weaknesses. A public-key cryptography as Elliptic-curve cryptography (ECC) is used in Bitcoin transactions. These transactions are authenticated using digital signatures, which are not random in nature compromising users' private access. Hash functions are other weaknesses of Bitcoin transactions. For example, if a hacker is able to access output $O_1$ from hashing input value $I_1$, he finds an output value $O_2$ from hashing input value $I_2$, hence a false data injection attack may occur. Mining malware refers to installing a malware on a mobile or other targeted device aiming to gain its computational capabilities. Software weaknesses comprises of runtime and synchronization aspects of Bitcoin transactions, which may lead to disclosure of users' private keys. Users' addresses are not certified in Bitcoin blockchain causing an identity fraud in transactions. Vulnerabilities related to consensus mechanism consists of double-spend and invalid transaction propagation attacks. Mining process threats consist of block withholding and pool hoping attacks. In block withholding attacks, the hacker connects to the mining pool without broadcasting any block for reducing the pool predicted revenue. The miner does not gain but everyone will lose. In the pool hoping attacks, the attacker will mine under high rates and leave pool at low-rate conditions [31, 45].

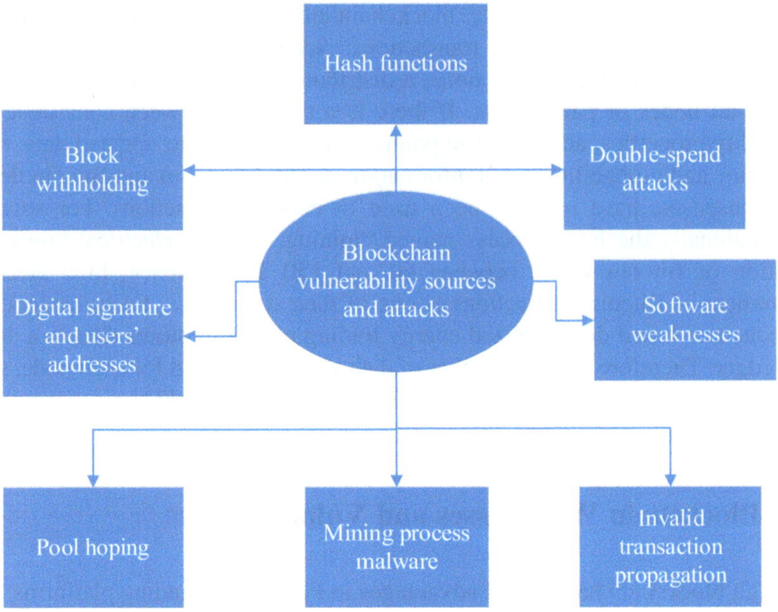


**Fig. 4.8** Bitcoin threats and vulnerabilities in P2P energy transactions

## 4.6 Concluding Remarks

The increasing growth of renewable energy sources in power systems necessitates the implementation of novel control strategies for optimal operation of power systems. Blockchain technology can be applied in various fields of smart grids, especially for peer-to-peer energy trading. Decreasing transaction costs, improving system resiliency, increasing system security, and guaranteeing privacy are among the most important benefits of blockchain technologies. Smart grids equipped with blockchain technology play a critical role in providing flexible and innovative solutions in the management and control of modern power systems. The concept of P2P energy transactions has also been successfully utilized in many cases around the world.

In this chapter, the application of blockchain technologies in smart energy grids with a focus on P2P energy trading was investigated. The general structure of the blockchain and its components were introduced and their relationships and the way they interact were explained. In addition, the main benefits as well as the most important challenges to deal with security and privacy issues were discussed.

## References


1. Adeyemi, A., et al. (2020). Blockchain technology applications in power distribution systems. *The Electricity Journal, 33*(8), 106817.
2. Green, J., & Newman, P. (2017). Citizen utilities: The emerging power paradigm. *Energy Policy, 105*, 283–293.
3. Bauknecht, D., Funcke, S., & Vogel, M. (2020). Is small beautiful? A framework for assessing decentralised electricity systems. *Renewable and Sustainable Energy Reviews, 118*, 109543.
4. Hirsch, A., Parag, Y., & Guerrero, J. (2018). Microgrids: A review of technologies, key drivers, and outstanding issues. *Renewable and Sustainable Energy Reviews, 90*, 402–411.
5. Shahidehpour, M., Yan, M., Shikhar, P., Bahramirad, S., & Paaso, A. (2020). Blockchain for peer-to-peer transactive energy trading in networked microgrids: Providing an effective and decentralized strategy. *IEEE Electrification Magazine, 8*(4), 80–90.
6. Denny, F. I., & Dismukes, D. E. (2017). *Power system operations and electricity markets*. CRC Press.
7. Sousa, T., Soares, T., Pinson, P., Moret, F., Baroche, T., & Sorin, E. (2019). Peer-to-peer and community-based markets: A comprehensive review. *Renewable and Sustainable Energy Reviews, 104*, 367–378.
8. Liu, Y., Yu, Y., Gao, N., & Wu, F. (2020). A grid as smart as the internet. *Engineering, 6*(7), 778–788.
9. Andoni, M., et al. (2019). Blockchain technology in the energy sector: A systematic review of challenges and opportunities. *Renewable and Sustainable Energy Reviews, 100*, 143–174.
10. Dinesha, D. L., & Balachandra, P. (2022). Conceptualization of blockchain enabled interconnected smart microgrids. *Renewable and Sustainable Energy Reviews, 168*, 112848.
11. Li, Z., Bahramirad, S., Paaso, A., Yan, M., & Shahidehpour, M. (2019). Blockchain for decentralized transactive energy management system in networked microgrids. *The Electricity Journal, 32*(4), 58–72.
12. Benčić, F. M., & Žarko, I. P. (2018). Distributed ledger technology: Blockchain compared to directed acyclic graph. In *2018 IEEE 38th international conference on distributed computing systems (ICDCS)* (pp. 1569–1570). IEEE.
13. Hu, J., Harmsen, R., Crijns-Graus, W., Worrell, E., & van den Broek, M. (2018). Identifying barriers to large-scale integration of variable renewable electricity into the electricity market: A literature review of market design. *Renewable and Sustainable Energy Reviews, 81*, 2181–2195.
14. Meeuw, A., et al. (2020). Implementing a blockchain-based local energy market: Insights on communication and scalability. *Computer Communications, 160*, 158–171.
15. Münsing, E., Mather, J., & Moura, S. (2017). Blockchains for decentralized optimization of energy resources in microgrid networks. In *2017 IEEE conference on control technology and applications (CCTA)* (pp. 2164–2171). IEEE.
16. Szabo, N. (2016). *Smart contracts: 12 use cases for business & beyond a technology, legal & regulatory introduction*, Chamber of Digital Commerce. https://digitalchamber.org/chamber-of-digitalcommerce-releases-smart-contracts-white-paper
17. Kaderali, F. (2007). Foundations and applications of cryptology.
18. Mingxiao, D., Xiaofeng, M., Zhe, Z., Xiangwei, W., & Qijun, C. (2017). A review on consensus algorithm of blockchain. In *2017 IEEE international conference on systems, man, and cybernetics (SMC)* (pp. 2567–2572). IEEE.
19. Macrinici, D., Cartofeanu, C., & Gao, S. (2018). Smart contract applications within blockchain technology: A systematic mapping study. *Telematics and Informatics, 35*(8), 2337–2354.
20. Berdik, D., Otoum, S., Schmidt, N., Porter, D., & Jararweh, Y. (2021). A survey on blockchain for information systems management and security. *Information Processing & Management, 58*(1), 102397.
21. Mohanta, B. K., Jena, D., Panda, S. S., & Sobhanayak, S. (2019). Blockchain technology: A survey on applications and security privacy challenges. *Internet of Things, 8*, 100107.

22. Sundararajan, A., Khan, T., Moghadasi, A., & Sarwat, A. I. (2019). Survey on synchrophasor data quality and cybersecurity challenges, and evaluation of their interdependencies. *Journal of Modern Power Systems and Clean Energy, 7*(3), 449–467.
23. Zhang, R., Xue, R., & Liu, L. (2019). Security and privacy on blockchain. *ACM Computing Surveys (CSUR), 52*(3), 1–34.
24. Mollah, M. B., et al. (2020). Blockchain for future smart grid: A comprehensive survey. *IEEE Internet of Things Journal, 8*(1), 18–43.
25. Aghapour, R., Zeraati, M., Jabari, F., Sheibani, M., & Arasteh, H. (2022). Cybersecurity and data privacy issues of electric vehicles smart charging in smart microgrids. In *Electric vehicle integration via smart charging* (pp. 85–110). Springer.
26. Musleh, A. S., Yao, G., & Muyeen, S. (2019). Blockchain applications in smart grid–review and frameworks. *IEEE Access, 7*, 86746–86757.
27. Fontana, P., Diirr, B., & Cappelli, C. (2018). Transparency challenges in Blockchain. *EGOV-CeDEM-ePart 2018*, p. 193.
28. Esmat, A., de Vos, M., Ghiassi-Farrokhfal, Y., Palensky, P., & Epema, D. (2021). A novel decentralized platform for peer-to-peer energy trading market with blockchain technology. *Applied Energy, 282*, 116123.
29. Afzal, M., et al. (2022). Role of blockchain technology in transactive energy market: A review. *Sustainable Energy Technologies and Assessments, 53*, 102646.
30. Brilliantova, V., & Thurner, T. W. (2019). Blockchain and the future of energy. *Technology in Society, 57*, 38–45.
31. Wang, T., Hua, H., Wei, Z., & Cao, J. (2022). Challenges of blockchain in new generation energy systems and future outlooks. *International Journal of Electrical Power & Energy Systems, 135*, 107499.
32. Che, Z., Wang, Y., Zhao, J., Qiang, Y., Ma, Y., & Liu, J. (2019). A distributed energy trading authentication mechanism based on a consortium Blockchain. *Energies, 12*(15), 2878. [Online]. Available: https://www.mdpi.com/1996-1073/12/15/2878
33. Saxena, S., Farag, H., Brookson, A., Turesson, H., & Kim, H. (2019, 19-21 Nov). Design and field implementation of Blockchain based renewable energy trading in residential communities. In *2019 2nd international conference on smart grid and renewable energy (SGRE)* (pp. 1–6). https://doi.org/10.1109/SGRE46976.2019.9020672
34. Dorri, A., Hill, A., Kanhere, S., Jurdak, R., Luo, F., & Dong, Z. Y. (2019, 14-17 May). Peer-to-peer EnergyTrade: A distributed private energy trading platform. In *2019 IEEE international conference on Blockchain and cryptocurrency (ICBC)* (pp. 61–64). https://doi.org/10.1109/BLOC.2019.8751268
35. Faizan, M., Brenner, T., Foerster, F., Wittwer, C., & Koch, B. (2019). Decentralized bottom-up energy trading using Ethereum as a platform. *Journal of Energy Markets, 12*(2), 19.
36. Yu, Y., Guo, Y., Min, W., & Zeng, F. (2019). Trusted transactions in micro-grid based on blockchain. *Energies, 12*(10), 1952.
37. Gurmani, M. U., et al. (2019). Energy trading between prosumer and consumer in p2p network using blockchain. In *International conference on P2P, parallel, grid, cloud and internet computing* (pp. 875–886). Springer.
38. Mihaylov, M., Razo-Zapata, I., & Nowe, A. (2018). NRGcoin—A blockchain-based reward mechanism for both production and consumption of renewable energy. In *Transforming climate finance and green investment with blockchains* (pp. 111–131). Elsevier.
39. Castellanos, J. A. F., Coll-Mayor, D., & Notholt, J. A. (2017). Cryptocurrency as guarantees of origin: Simulating a green certificate market with the Ethereum Blockchain. In *2017 IEEE international conference on smart energy grid engineering (SEGE)* (pp. 367–372). IEEE.
40. Bokhari, S. T., Aftab, T., Nadir, I., & Bakhshi, T. (2019). Exploring blockchain-secured data validation in smart meter readings. In *2019 22nd international multitopic conference (INMIC)* (pp. 1–7). IEEE.
41. Onifade, O., Sorinolu, B., & Ibitoye, A. (2019). *Improved trust worthiness in electrical energy management using K-means augmented blockchain technology.*

42. Franke, L., Schletz, M., & Salomo, S. (2020). Designing a blockchain model for the Paris agreement's carbon market mechanism. *Sustainability, 12*(3), 1068.
43. Richardson, A., & Xu, J. (2020). Carbon trading with blockchain. In *Mathematical research for Blockchain economy* (pp. 105–124). Springer.
44. Suciu, G., et al. (2019). Securing the smart grid: A blockchain-based secure smart energy system. In *2019 54th international universities power engineering conference (UPEC)* (pp. 1–5). IEEE.
45. Wylde, V., et al. (2022). Cybersecurity, data privacy and blockchain: A review. *SN Computer Science, 3*(2), 1–12.

# Chapter 5
# Fully Decentralized and Competitive Hybrid Retail and Local Electricity Market Using Peer-to-Peer Energy Trading

**Mostafa Yaghoubi, Mehdi Mehdinejad, Mohammad Seyfi, Zbigniew Dziong, and Heidarali Shayanfar**

## 5.1 Introduction

### *5.1.1 Background and Motivation*

In traditional electricity networks, the power generated by generation units was delivered to the consumers unilaterally using a vertical and centralized operational structure through transmission and distribution systems [1]. The spread of communication infrastructures, i.e., the pervasiveness of smart meters, increased the penetration of distributed energy resources (DERs) on the consumers' side, leading to the development of distributed power generation worldwide. This contributed to the formation of a novel concept called "prosumers" at the distribution network level. Prosumers are defined as consumers who generate and store electricity at home using solar photovoltaic panels, electric vehicles, batteries, or other equipment [2]. The emergence of prosumers as new smart agents has transformed traditional electricity markets into prosumer-centric markets with P2P energy transaction capability in which the players are incentivized to engage in local energy trading with each other [3].

These markets have been redesigned extensively into various forms with different clearing approaches and considering the competitive characteristics [2]. Due to the

M. Yaghoubi · Z. Dziong
Department of Electrical Engineering, Ecole de Technologie Superieure (ETS), Montreal, QC, Canada
e-mail: mostafa.yaghoubi.1@ens.etsmtl.ca; zbigniew.dziong@etsmtl.ca

M. Mehdinejad (✉) · M. Seyfi · H. Shayanfar
Center of Excellence for Power Systems Automation and Operation, School of Electrical Engineering, Iran University of Science and Technology, Tehran, Iran

 
V. Vahidinasab, B. Mohammadi-Ivatloo (eds.), *Demand-Side Peer-to-Peer Energy Trading*, Green Energy and Technology, https://doi.org/10.1007/978-3-031-35233-1_5

low capacity of prosumers at the distribution network level, the focus on local energy transactions has been growing [4]. This type of design deprives the local players of the benefits of the wholesale market [5], as they have to rely on mediators to participate in the wholesale market [6]. Participation in the wholesale market via mediators prevents the formation of a fully competitive market for local players because, in a fully competitive market, each prosumer with any amount of production and consumption can freely and individually participate in the upstream and local markets [7]. In this regard, retailers act as mediators, trading energy with prosumers at a fixed price [8–10].

The clearing methods for electrical energy markets depend on assumptions, particular rules, market structure, and players' behaviors. Two general approaches are commonly used to clear these markets: centralized and decentralized. The former approach requires the information of all market players, which violates their privacy of decision-making information and conflicts with the competitive market model. In addition, the computational overhead of these methods is of concern. Conversely, in the latter, the market players solve their economic distribution problems individually and only exchange the price and quantity of transactions, which ensures the privacy protection of decision-making information of the market players. The most significant distributed methods for clearing P2P markets include the primal-dual subgradient [10–12], alternating direction method of multipliers (ADMM) [13], and consensus-based methods [14, 15].

### 5.1.2 Related Works

A literature review on the new market design is carried out from two aspects of P2P energy transactions and electricity retailers' place in these markets.

Reference [16] designed and implemented a decentralized P2P energy trading market, considering the power losses and the network utilization fee at the transmission level. In the proposed model, the sellers are considered price makers, and their transactions are conducted at the same price. A P2P market for energy buildings is established in [17], in which a non-cooperative sharing game is used to determine P2P transactions. In the non-cooperative sharing game, each energy building offers its supply/demand amount at the considered price under constraints of community energy and payment balance. Authors in [18] developed a P2P market for energy trading between smart buildings equipped with electric vehicle charging stations and solar photovoltaic panels. In the proposed market model, they introduced a dynamic pricing mechanism to encourage market players to purchase from smart buildings and participate in P2P energy trading with electric vehicles. The purchase price from these players was considered lower than the upstream network price. A P2P market was presented in [14] using the multi-bilateral economic dispatch concept and cleared by the relaxed consensus innovation (RCI) approach. The authors claimed that the proposed approach preserves the decentralized nature of the designed market. Reference [19] developed an energy trading framework between prosumers equipped with storage that can participate in the upstream market. A distributed energy management model using robust scheduling was presented for prosumers, considering the upstream market price uncertainty. The proposed market was cleared in a decentralized manner using the Fast ADMM approach. An energy token market for prosumers equipped with blockchain-based two-way electric meters is proposed in [20]. In this model, a novel concept called demurrage is used to incentivize local transactions and prevent the hoarding of energy-backed tokens.

The interactions between small consumers and retailers encourage competition on the demand side and promote the prosumers' benefits from the upstream market. Thus, retailers can participate in the wholesale market as buyers/sellers while trading energy with prosumers via bilateral contracts. Previous studies on market design and electricity retail management models have not addressed this concept. For instance, authors in [21] designed an energy market for price-maker retailers without considering the P2P concept, where the retailers adopted a Stackelberg-based pricing mechanism. The proposed model is a bi-level framework in which the upper level consists of a price-maker retailer as the leader seeking to maximize its profit, while the lower level involves four followers, including three consumers with flexible loads, whose objective function is to minimize the energy consumption cost, and one pooled market aiming for welfare maximization. An energy management model of an integrated retailer in multi-energy systems is proposed in [22]. The retailer is assumed as an energy hub attempting to maximize its profit by managing multiple types of energy, including electricity, natural gas, heat, and cold. Reference [23] presented a bilateral model of energy trading between consumers and a retailer in which the consumers equipped with smart grid technology from the retailer's viewpoint participate in short-term demand response. In the proposed model, the consumers offer their short-term DR curves to the retailer to reduce or increase their energy consumption for a given period based on the energy price. An aggregation framework for prosumers based on transactive energy is introduced in [5], where prosumers aggregate independently without relying on a real central entity and trade as a retailer in the wholesale market and with end-users in the retail market. The proposed model was solved using an inner-outer iteration approach.

In this chapter, a P2P energy market is designed for bilateral and multi-lateral interactions between prosumers and active retailers in the geographical region of prosumers. In this market, each prosumer with energy deficiency (as the buyer) purchases its required energy based on the price bids of retailers and local sellers (prosumers with excess energy). Based on its unique economic policies, each smart agent (prosumer and retailer) has the right to choose whether to participate in the proposed bilateral or multilateral market in the distribution network. On the other hand, retailers can participate in the upstream market. The proposed decentralized market is cleared using a novel approach called primal-dual subgradient. Rather than using the interior-point method to find primary variables, this scheme uses the first-order method, which requires less time per iteration. In addition, the convergence coefficients of sub-problems can be generalized to any problem with any number of players (buyers, sellers, retailers, and so forth). The contributions of this article are summarized as follows:

- Design of a fully competitive P2P market, including prosumers and retailers
- Development of a decentralized approach called the primal-dual sub-gradient method to clear the proposed decentralized market while protecting the privacy of players' decision-making information.

The rest of this chapter is organized as follows:

Section 5.2: Conceptual and mathematical modeling of the proposed market and the proposed market clearing mechanism
Section 5.3: Numerical analysis of the proposed market model
Section 5.4: The results of this chapter are presented.

## 5.2 P2P Energy Trading Platform

### *5.2.1 Description of Market Structure, Assumptions, and Players*

The proposed market model consists of three types of players. The first group is retailers active in the geographical region of the prosumers. The second and third groups include prosumers who are assumed to have higher/lower generation than consumption during the scheduling period, acting as sellers/buyers in the proposed market. During the scheduling period, all groups, with the assumption of establishing telecommunication connections, negotiate individually over the energy price and quantity and reach an agreement by satisfying the stopping condition. In other words, the proposed market is cleared through P2P interactions among all market players. Notably, the second and third groups of market players are assumed to have flexible generation and consumption. From the time perspective, the proposed market is a forward market. The market-clearing is performed for a time slot (one hour) and is extendable to multiple time slots.

### *5.2.2 The Mathematical Modeling of the Proposed Market*

The proposed market comprises $\mathcal{N}$ players, including $\mathcal{N}_\mathrm{S} = \{1, \ldots, N_\mathrm{S}\}$ local sellers, $\mathcal{N}_\mathrm{B} = \{1, \ldots, N_\mathrm{B}\}$ local buyers, and $\mathcal{N}_\mathrm{R} = \{1, \ldots, N_\mathrm{R}\}$ active retailers in the studied district, such that $\mathcal{N}_\mathrm{B} \cap \mathcal{N}_\mathrm{S} \cap \mathcal{N}_\mathrm{R} = \varnothing$.

A. *The buyer's model*

In the proposed market model, the buyers have flexible loads and aim to maximize their welfare by adjusting their demands according to the energy price. The following relations express the objective function and constraints corresponding to each buyer:

$$\max_{y} \quad \mathrm{WB}_j = U_j(y_j) - \sum_{r \in \mathcal{N}_\mathrm{R}} \lambda_{rj} y_{jr} - \sum_{i \in \mathcal{N}_\mathrm{S}} \lambda_{ij} y_{ji} \tag{5.1}$$

Subject to

$$y_j = \sum_i y_{ji} + \sum_r y_{jr} \tag{5.2}$$

$$y_j^{\min} \leq y_j \leq y_j^{\max} : \overline{\mu}_j, \underline{\mu}_j \tag{5.3}$$

$$U(y_j) = \begin{cases} \omega_j y_j - \delta_j y_j^2 & y_j < \dfrac{\omega_j}{2\delta_j} \\ \dfrac{\omega_j^2}{2\delta_j} & y_j \geq \dfrac{\omega_j}{2\delta_j} \end{cases} \tag{5.4}$$

The objective function of these players comprises two terms. The first term models the buyers' satisfaction level and response to energy price variations using the utility function concept. Relation (5.4) describes the utility function for buyer $j$ [24]. The second term describes the cost of energy purchases from retailers and local sellers. Relation (5.2) gives the total energy purchased by buyer $\boldsymbol{j}$ from local sellers and retailers. As shown in relation (5.3), the total energy purchase is constrained by minimum and maximum demands (flexible load). Based on the presented model, the parameters and decision variables for buyer $\boldsymbol{j}$ are as follows:

- $\omega_j$ and $\delta_j$ are known positive parameters in the utility function, which contain the private information of buyer $\boldsymbol{j}$ for decision-making in the proposed market.
- $y_j^{\min}$ and $y_j^{\max}$ are positive parameters in the energy purchase constraint and represent the minimum and maximum required energy for buyer $\boldsymbol{j}$.
- $y_{jr}$ and $\lambda_{rj}$ are continuous positive variables, modeling the purchased energy and its corresponding price in trade between retailer $\boldsymbol{r}$ and buyer $\boldsymbol{j}$. $y_{jr}$ is determined by the buyer and $\lambda_{rj}$ by the retailer.
- $y_{ji}$ and $\lambda_{ij}$ are continuous positive variables, modeling the purchased energy and its corresponding price in trade between seller $\boldsymbol{i}$ and buyer $\boldsymbol{j}$. $y_{ji}$ is determined by the buyer and $\lambda_{ij}$ by the local seller.
- $\overline{\mu}_j$ and $\underline{\mu}_j$ are positive variables called Lagrangian coefficients, associated with minimum and maximum energy purchase constraints (5.3).

B. *Seller model*

The local sellers seek to maximize their welfare by selling their surplus energy to buyers and local retailers. These players adjust their excess energy generation based on the local market price. The following relations express the objective function and constraints for local seller $\boldsymbol{i}$:

$$\max_{x} \quad \text{WS}_i = \sum_{j \in N_\text{B}} \lambda_{ij} x_{ij} + \sum_{r \in N_\text{R}} \lambda_{ir} x_{ir} - C(x_i) \tag{5.5}$$

Subject to

$$x_i = \sum_j x_{ij} + \sum_r x_{ir} \tag{5.6}$$

$$x_i^{\min} \le x_i \le x_i^{\max} \qquad : \overline{\mu}_i, \underline{\mu}_i \tag{5.7}$$

$$C(x_i) = \alpha_i x_i^2 + \beta_i x_i + \gamma_i \tag{5.8}$$

The first two terms in relation (5.5) indicate the income from selling energy to consumer ***j*** and retailer ***r***. The third term of the seller's objective function models the energy generation cost, as shown in relation (5.8) [25]. Relation (5.6) yields the total energy sold to all energy peers and is constrained by relation (5.7). The parameters and decision variables in the mathematical model of seller ***i*** are as follows:

- $\alpha_i$, $\beta_i$, and $\gamma_i$ are known positive parameters in the cost function, which contain the private information of seller ***i*** for decision-making in the proposed market.
- $x_i^{\min}$ and $x_i^{\max}$ are positive parameters in the energy generation constraint and represent the minimum and maximum energy generation for seller ***i***.
- $x_{ir}$ and $\lambda_{ir}$ are continuous positive variables, modeling the sold energy and its corresponding price in the trade between seller ***i*** and retailer ***r***. $x_{ir}$ and $\lambda_{ir}$ are both determined by the seller.
- $x_{ij}$ and $\lambda_{ij}$ are continuous positive variables, modeling the sold energy and its corresponding price in the trade between seller ***i*** and buyer ***j***. $x_{ij}$ and $\lambda_{ij}$ are both determined by the seller.
- $\overline{\mu}_i$ and $\underline{\mu}_i$ denote the Lagrangian coefficients associated with minimum and maximum energy generation constraints (5.7).

C. *Retailer's model*

The retailers are profit-centric units that maximize their profit by simultaneous participation in local and wholesale markets. These peers deliver the generated energy by their generators to buyers and the upstream network and can purchase energy from local sellers and the upstream market. Therefore, these entities can play both buyer and seller roles during the scheduling period.

$$\max_{z} \quad R_r = \lambda^{\text{DA}} z_{r,g} + \sum_j \lambda_{rj} z_{rj} - \sum_i \lambda_{ir} z_{ri} - C(z_r) \tag{5.9}$$

Subject to

$$z_r = z_{r,g} + \sum_j z_{rj} - \sum_i z_{ri} \tag{5.10}$$

$$z_r^{\min} \le z_r \le z_r^{\max} : \overline{\mu}_r, \underline{\mu}_r \tag{5.11}$$

$$C(z_r) = \alpha_r z_r^2 + \beta_r z_r + \gamma_r \tag{5.12}$$

$$z_{r,g} \text{ is free}$$

Relation (5.9) models the income of retailer ***r***, which consists of four terms. The first term in this relation indicates the income/cost of energy trading with the upstream market. For a positive $z_{r,\,g}$, the retailers seek to sell energy to the upstream market, and this term is considered a part of the retailer's income. However, if $z_{r,\,g}$ is negative, the retailer seeks to purchase energy from the upstream market, and this term is a part of the retailer's costs. The second term in the objective function of retailer ***r*** indicates the income from selling energy to the local consumer ***i***. The third one is the cost of energy purchase from the self-generator ***i***, and the last term is the cost of energy generation by the self-generation unit. The last term of the retailer income is modeled by relation (5.12). Relation (5.10) describes the amount of energy production by the self-generation unit of retailer ***r***, and the amount of energy generation is constrained by relation (5.11). The parameters and decision variables of the mathematical model of retailer ***r*** are as follows:

- $\alpha_r$, $\beta_r$, and $\gamma_r$ are positive parameters in the cost function, which contain private information unique to retailer ***r*** and depend on its generation type and future risks.
- $z_{r,\,g}$ is a continuous free-sign variable that shows the traded energy with the upstream network.
- $\lambda^{\mathrm{DA}}$ is a positive parameter that shows the upstream market price.
- $z_r$ is a continuous positive variable indicating the self-generation of retailer ***r*** so that $z_r > 0$ means that the retailer has utilized its generators to participate in the market.
- $\overline{\mu}_r$ and $\underline{\mu}_r$ denote the Lagrangian coefficients related to the minimum and maximum energy generation constraint for the retailer (5.11).

D. *Optimization problem with centralized clearing approach*

The final objective function of the problem in the centralized approach is obtained from the summation of the total welfare of prosumers in the local market and the total income of active retailers in this market.

$$\max_{x,y,z} \left( \sum_{j=1}^{\mathcal{N}_B} U(y_j) - \sum_{i=1}^{\mathcal{N}_S} C(x_i) + \sum_{r=1}^{\mathcal{N}_R} \left( \lambda^{\mathrm{DA}} z_{r,g} - C(z_r) \right) \right) \tag{5.13}$$

s.t.

$$x_{ij} = y_{ji} : \lambda_{ij} \tag{5.14}$$

$$x_{ir} = z_{ri} : \lambda_{ir} \tag{5.15}$$

$$z_{rj} = y_{jr} : \lambda_{rj} \tag{5.16}$$

$$\text{Constraints (5.2) to (5.4), and (5.6) to (5.8), and (5.10) to (5.12)}$$

Relation (5.13) gives the final objective function of the centralized approach, which is obtained from the sum of relations (5.1), (5.5), and (5.9) for all market players. It determines the demand of all local buyers ($y$), the generation of all local sellers ($x$), and the generated and purchased power of all retailers ($z$). Relations (5.14), (5.15), and (5.16) model the supply and demand constraints of the energy bid between individual prosumers and retailers in P2P transactions, which are also known as coupled constraints or power balance constraints. The dual of these constraints ($\lambda_{ij}$, $\lambda_{ir}$, $\lambda_{rj}$) describes the agreed price for each transaction between the players (marginal cost).

E. *Design of decentralized market clearing algorithm*

A centralized approach for solving the proposed market requires a supervisory node aware of all market players' characteristics. Such access to the decision parameters of the players puts their privacy at risk and affects the market competitiveness. The primal-dual sub-gradient decentralized approach in this chapter allows each agent (player) to perform its calculations locally.

$$
\begin{aligned}
&\mathcal{L}\left(x_{ij}, y_{ji}, \lambda_{ij}, x_{ir}, z_{ri}, \lambda_{ir}, z_{rj}, y_{jr}, \lambda_{rj}, z_{r,g}, \overline{\mu}_j, \underline{\mu}_j, \overline{\mu}_r, \underline{\mu}_r, \overline{\mu}_i, \underline{\mu}_i\right) = \sum_{j=1}^{N_{\mathrm{B}}} U_j\left(y_j\right) \\
&\quad + \sum_{r=1}^{N_{\mathrm{R}}} \lambda^{\mathrm{DA}} z_{r,g} - \sum_{i=1}^{N_{\mathrm{S}}} C_i(x_i) - \sum_{r=1}^{N_{\mathrm{R}}} C_r(z_r) + \sum_{i=1}^{N_{\mathrm{B}}} \sum_{j=1}^{N_{\mathrm{S}}} \lambda_{ij}\left(x_{ij} - y_{ji}\right) + \sum_{i=1}^{N_{\mathrm{S}}} \\
&\quad \times \sum_{r=1}^{N_{\mathrm{R}}} \lambda_{ir}\left(x_{ir} - z_{ri}\right) + \sum_{r=1}^{N_{\mathrm{R}}} \sum_{j=1}^{N_{\mathrm{B}}} \lambda_{rj}\left(z_{rj} - y_{jr}\right) - \sum_{j=1}^{N_{\mathrm{B}}} \overline{\mu}_j\left(y_j - y_j^{\max}\right) \\
&\quad - \sum_{j=1}^{N_{\mathrm{B}}} \underline{\mu}_j\left(y_j^{\min} - y_j\right) - \sum_{r=1}^{N_{\mathrm{R}}} \overline{\mu}_r\left(z_r - z_r^{\max}\right) - \sum_{r=1}^{N_{\mathrm{R}}} \underline{\mu}_r\left(z_r^{\min} - z_r\right) \\
&\quad - \sum_{i=1}^{N_{\mathrm{S}}} \overline{\mu}_i\left(x_i - x_i^{\max}\right) - \sum_{i=1}^{N_{\mathrm{S}}} \underline{\mu}_i\left(x_i^{\min} - x_i\right)
\end{aligned}
\tag{5.17}
$$

$$
\begin{aligned}
&\Upsilon\left(\lambda_{ij}, \lambda_{ir}, \lambda_{rj}, \overline{\mu}_j, \underline{\mu}_j, \overline{\mu}_r, \underline{\mu}_r, \overline{\mu}_i, \underline{\mu}_i\right) \\
&\quad = \sup_{x,y,z} \mathcal{L}\left(x_{ij}, y_{ji}, \lambda_{ij}, x_{ir}, z_{ri}, \lambda_{ir}, z_{rj}, y_{jr}, \lambda_{rj}, z_{r,g}, \overline{\mu}_j, \underline{\mu}_j, \overline{\mu}_r, \underline{\mu}_r, \overline{\mu}_i, \underline{\mu}_i\right) \\
&\quad = \sum_{i \in \mathcal{N}_{\mathrm{S}}} \underset{x_i^{\min} \le x_i \le x_i^{\max}}{\arg\max}\ \mathfrak{S}_i\left(\lambda_{ij}, \lambda_{ir}, \overline{\mu}_i, \underline{\mu}_i\right) + \sum_{j \in \mathcal{N}_{\mathrm{B}}} \underset{y_j^{\min} \le y_j \le y_j^{\max}}{\arg\max}\ \mathfrak{B}_j\left(\lambda_{ij}, \lambda_{rj}, \overline{\mu}_j, \underline{\mu}_j\right) \\
&\quad + \sum_{r \in \mathcal{N}_{\mathrm{R}}} \underset{z_r^{\min} \le z_r \le z_r^{\max}}{\arg\max}\ \mathcal{R}_r\left(\lambda_{ir}, \lambda_{rj}, \overline{\mu}_j, \underline{\mu}_j\right) + \sum_{i \in \mathcal{N}_{\mathrm{S}}} \left(\underline{\mu}_i x_i^{\min} - \overline{\mu}_i x_i^{\max}\right) \\
&\quad + \sum_{j \in \mathcal{N}_{\mathrm{B}}} \left(\underline{\mu}_j y_j^{\min} - \overline{\mu}_j y_j^{\max}\right) + \sum_{r \in \mathcal{N}_{\mathrm{R}}} \left(\underline{\mu}_r z_r^{\min} - \overline{\mu}_r z_r^{\max}\right)
\end{aligned}
\tag{5.18}
$$

$$\mathfrak{s}_i\left(\lambda_{ij},\lambda_{ir},\overline{\mu}_i,\underline{\mu}_i\right)\triangleq\sum_{j\in N_{\mathrm{B}}}\lambda_{ij}x_{ij}+\sum_{r\in N_{\mathrm{R}}}\lambda_{ir}x_{ir}-C(x_i)+\left(\underline{\mu}_i-\overline{\mu}_i\right)x_i \tag{5.19}$$

$$\mathfrak{B}_j\left(\lambda_{ij},\lambda_{rj},\overline{\mu}_j,\underline{\mu}_j\right)\triangleq U_j(y_j)-\sum_{r\in\mathcal{N}_{\mathrm{R}}}\lambda_{rj}y_{jr}-\sum_{i\in\mathcal{N}_{\mathrm{S}}}\lambda_{ij}y_{ji}+\left(\underline{\mu}_j-\overline{\mu}_j\right)y_j \tag{5.20}$$

$$\mathcal{R}_r\left(\lambda_{ir},\lambda_{rj},\overline{\mu}_j,\underline{\mu}_j\right)\triangleq\lambda^{\mathrm{DA}}z_{r,g}+\sum_{j}\lambda_{rj}z_{rj}-\sum_{i}\lambda_{ir}z_{ri}-C(z_r)+\left(\underline{\mu}_r-\overline{\mu}_r\right)z_r \tag{5.21}$$

$$\underset{\lambda_{ij},\lambda_{ir},\lambda_{rj},\overline{\mu}_j,\underline{\mu}_j,\overline{\mu}_r,\underline{\mu}_r,\overline{\mu}_i,\underline{\mu}_i}{\arg\max}\Upsilon\left(\lambda_{ij},\lambda_{ir},\lambda_{rj},\overline{\mu}_j,\underline{\mu}_j,\overline{\mu}_r,\underline{\mu}_r,\overline{\mu}_i,\underline{\mu}_i\right)$$

Subject to

$$\begin{gathered}\lambda_{ij},\lambda_{ir},\lambda_{rj},\overline{\mu}_j,\underline{\mu}_j,\overline{\mu}_r,\underline{\mu}_r,\overline{\mu}_i,\underline{\mu}_i\geq 0\\ \forall i\in N_{\mathrm{S}},\forall j\in N_{\mathrm{B}},\forall r\in N_{\mathrm{R}}\end{gathered} \tag{5.22}$$

$$\mu^{k+1}=\left[\mu^{k+1}-\tau\nabla_{\mu}\Upsilon\right]^{+} \tag{5.23}$$

$$x_i^k=\left[\left(\mathfrak{s}_i'\right)^{-1}\left(\lambda_{ij}^k,\lambda_{ir}^k\right)\right]_{x_i^{\min}}^{x_i^{\max}} \tag{5.24}$$

$$y_j^k=\left[\left(\mathfrak{B}_j'\right)^{-1}\left(\lambda_{ij}^k,\lambda_{rj}^k\right)\right]_{y_j^{\min}}^{y_j^{\max}} \tag{5.25}$$

$$z_r^k=\left[\left(\mathcal{R}_r'\right)^{-1}\left(\lambda_{ir}^k,\lambda_{rj}^k,\lambda^{\mathrm{DA}}\right)\right]_{z_r^{\min}}^{z_r^{\max}} \tag{5.26}$$

$$x_{ij}^{k+1}=\left[\frac{\lambda_{ij}^{k+1}-\overline{\mu}_i^{k+1}+\underline{\mu}_i^{k+1}-\beta_i}{2\alpha_i}\right]_{x_i^{\min}}^{x_i^{\max}} \tag{5.27}$$

$$x_{ir}^{k+1}=\left[\frac{\lambda_{ir}^{k+1}-\overline{\mu}_i^{k+1}+\underline{\mu}_i^{k+1}-\beta_i}{2\alpha_i}\right]_{x_i^{\min}}^{x_i^{\max}} \tag{5.28}$$

$$y_{ji}^{k+1}=\left[\frac{\omega_j-\lambda_{ij}^{k+1}-\overline{\mu}_j^{k+1}+\underline{\mu}_j^{k+1}}{2\delta_j}\right]_{y_j^{\min}}^{y_j^{\max}} \tag{5.29}$$

$$y_{jr}^{k+1}=\left[\frac{\omega_j-\lambda_{rj}^{k+1}-\overline{\mu}_j^{k+1}+\underline{\mu}_j^{k+1}}{2\delta_j}\right]_{y_j^{\min}}^{y_j^{\max}} \tag{5.30}$$

$$z_{ri}^{k+1}=\left[\frac{-\lambda_{ir}^{k+1}+\overline{\mu}_r^{k+1}-\underline{\mu}_r^{k+1}+\beta_r}{2\alpha_r}\right]_{z_r^{\min}}^{z_r^{\max}} \tag{5.31}$$

$$z_{rj}^{k+1} = \left[\frac{\lambda_{rj}^{k+1} - \overline{\mu}_r^{k+1} + \underline{\mu}_r^{k+1} - \beta_r}{2\alpha_r}\right]_{z_r^{\min}}^{z_r^{\max}} \tag{5.32}$$

$$z_{r,g}^{k} = \left[\frac{\lambda^{\mathrm{DA}} - \overline{\mu}_r^{k+1} + \underline{\mu}_r^{k+1} - \beta_r}{2\alpha_r}\right]^{\pm} \tag{5.33}$$

$$\begin{cases} \left|\lambda_{ij}^{k+1} - \lambda_{ij}^{k}\right| \leq \epsilon, \left|\lambda_{ir}^{k+1} - \lambda_{ir}^{k}\right| \leq \epsilon \\ \left|\overline{\mu}_i^{k+1} - \overline{\mu}_i^{k}\right| \leq \epsilon, \left|\underline{\mu}_i^{k+1} - \underline{\mu}_i^{k}\right| \leq \epsilon \end{cases} \tag{5.34}$$

$$\left|\overline{\mu}_j^{k+1} - \overline{\mu}_j^{k}\right| \leq \epsilon, \left|\underline{\mu}_j^{k+1} - \underline{\mu}_j^{k}\right| \leq \epsilon \tag{5.35}$$

$$\left|\lambda_{rj}^{k+1} - \lambda_{rj}^{k}\right| \leq \epsilon, \left|\overline{\mu}_r^{k+1} - \overline{\mu}_r^{k}\right| \leq \epsilon, \left|\underline{\mu}_r^{k+1} - \underline{\mu}_r^{k}\right| \leq \epsilon \tag{5.36}$$

The proposed centralized model is decomposed into several secondary sub-problems based on the dual decomposition principle to solve the proposed model in a decentralized manner [26]. Thus, a secondary problem is developed for each agent. The following steps should be taken to solve the model using the proposed decentralized approach:

*Step 1:* By relaxing the global and local constraints, the Lagrangian of the proposed model in the centralized approach is achieved as in relation (5.17), based on the Lagrangian multipliers. This operation converts the proposed model into an unconstrained model.

*Step 2:* The dual function is obtained from Lagrangian supremum on variables of the main problem, as shown in relation (5.18), where:

- $\mathfrak{s}_i\left(\lambda_{ij}, \lambda_{ir}, \overline{\mu}_i, \underline{\mu}_i\right)$ defined in relation (5.19) is a sub-problem maximized by seller ***i***.
- $\mathfrak{B}_j\left(\lambda_{ij}, \lambda_{rj}, \overline{\mu}_j, \underline{\mu}_j\right)$ defined in relation (5.20) is a sub-problem maximized by buyer ***j***.
- $\mathcal{R}_r\left(\lambda_{ir}, \lambda_{rj}, \overline{\mu}_j, \underline{\mu}_j\right)$ defined in relation (5.21) is a sub-problem maximized by retailer ***r***.

*Step 3:* The dual function is rewritten as relation (5.22) based on the Lagrangian multipliers.

*Step 4:* The Lagrangian multipliers are calculated iteratively using the sub-gradient projection method according to relation (5.23). From this relation, the Lagrangian multipliers are updated in the opposite direction of the dual function sub-gradient $(-\nabla_\mu \Upsilon)$.

- $\tau$ is a positive input parameter called the convergence step size that guarantees model convergence.
- $[.]^+$ is an operator, which represents $\max(0, .)$.

*Step 5:* At each iteration $\boldsymbol{k}$, primary variables are calculated using relations (5.24), (5.25), and (5.26), which should be solved using the first-order and sub-gradient projection methods. These relations can be extended to relations (5.27), (5.28), (5.29), (5.30), (5.31), (5.32), and (5.33) using the inverse gradient and its projection on the feasibility space to achieve the target energy setpoint for all players.

*Step 6:* Stopping criteria of the decentralized approach are adjusted as a function of the difference between two consecutive iterations, as shown in relations (5.34), (5.35), and (5.36). $\epsilon$ is a small positive number.

## 5.3 Simulation

This section presents the numerical studies to evaluate the performance of the designed P2P market model and the effectiveness of the proposed decentralized approach.

### 5.3.1 *Test System*

The studied system consists of 7 prosumers, including 3 buyers, four sellers, and 3 active retailers in a small residential distribution network. All parameters in the mathematical model of prosumers are derived from [10]. Cost function parameters, minimum and maximum self-generation capacities for each retailer, are generated randomly and taken as $\alpha_r = [0.09, 0.1]$, $\beta_r = [8, 15]$, $z_r^{\max} = [6, 9]$, and $z_r^{\min} = 0$. The stopping criterion for the proposed decentralized algorithm is assumed as 0.001, and the convergence step is 0.01.

Three case studies are considered in this chapter to assess the proposed model and its clearing approach:

- *Case study 1:* P2P energy trading without the presence of retailers
- *Case study 2:* P2P energy trading with the presence of retailers and fixed wholesale price
- *Case study 3:* P2P energy trading with the presence of retailers and variable wholesale price

To validate the presented decentralized clearing approach, the proposed market model is cleared using the centralized approach in the GAMS software, and the results of the two approaches are compared.

### 5.3.2 Case Study 1: P2P Energy Trading Without the Presence of Retailers

This case study assumes that no retailer is active in the prosumers' district and prosumers only carry out P2P energy trading with each other. Given the problem dimensions, considering all negotiations and transactions between buyers and sellers will result in repetitive and redundant information. Thus, to avoid further complexity, we only assess the negotiations between buyer 2 and all sellers in this case study, as illustrated in Fig. 5.1. According to this figure, sellers offer lower prices during the first 15 iterations, and thus buyer 2 requests to purchase energy from all sellers. After 100 iterations, the prices tend to become more realistic and consistent with the objectives of sellers and the local market, with buyer 2 also offering more realistic energy purchase requests to the sellers. For instance, seller 4, who offers the highest price, receives no energy purchase requests, whereas seller 1, with the lowest bid, receives the most requests for energy purchases. Figure 5.1 demonstrates the convergence of the supply and demand offers and, consequently, the proposed algorithm. The local market has been cleared in 1.2 s and 498 iterations in this case study.

Table 5.1 presents the welfare of all prosumers and the overall welfare of the proposed market. As can be seen, seller four and buyer one have achieved the lowest welfare from P2P trading. This table also compares the centralized and decentralized approaches regarding the final welfare of the whole market, suggesting that the proposed decentralized solution is extremely close to the global optimum achieved by the centralized approach. Thus, the presented algorithm can clear the proposed market in a fully decentralized manner without interference from a central entity while ensuring the players' privacy.

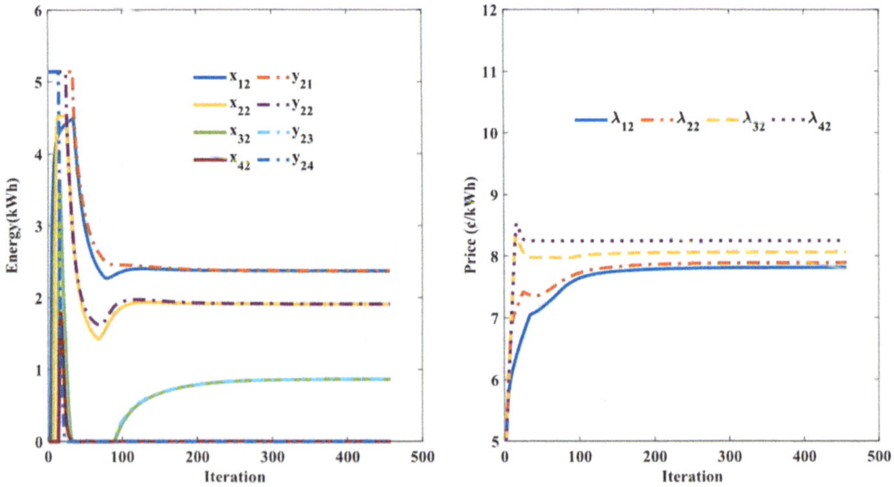


**Fig. 5.1** The bilateral negotiations between buyer two and all sellers

**Table 5.1** The welfare of local players and comparison of the proposed decentralized and centralized approaches in P2P trading

| Welfare (¢) | $WS_1$ | $WS_2$ | $WS_3$ | $WS_4$ | $WB_1$ | $WB_2$ | $WB_3$ | Total | |
|---|---|---|---|---|---|---|---|---|---|
| | | | | | | | | Proposed method | Centralized |
| P2P Market | 16.24 | 7.19 | 0.08 | 0.00 | 0.84 | 16.24 | 16.60 | 57.20 | 57.2240 |

### 5.3.3 *Case Study 2: P2P Energy Trading with the Presence of Retailers and Fixed Wholesale Price*

This case study evaluates the effect of retailers' presence in the proposed market. Retailers are considered energy peers for prosumers that can trade energy with the upstream market at a fixed price. Interaction with retailers allows prosumers to enjoy the upstream market benefits indirectly. Thus, for this case study, each prosumer can trade energy (buying or selling) with its local prosumer as well as the retailer. Each retailer can trade energy with the upstream market, in addition to local prosumers (buying/selling). The wholesale market price is taken as 7.98 cents. Given the problem dimensions, Fig. 5.2 depicts the transactions between buyer 1 with all buyer prosumers and retailers active in their geographical region. The negotiations over the energy amount and its equivalent price are plotted in parts (a) and (b) of this figure, respectively. From part (a), buyer 1 has purchased the highest amount of energy from the local seller 1 and no energy from seller 4. The negotiation process over the first 20 iterations suggests that buyer one has been initially unwilling to purchase energy from seller four given its high price bid and instead has been focused on purchasing energy from the sellers offering lower prices. Among the retailers, retailer 1 and retailer 3 have the maximum and minimum shares from the energy transactions, respectively. Regarding the negotiation process with retailers, there has been no negotiation between buyer 1 and the retailers over purchasing energy during the first 20 iterations. Buyer one negotiated with retailer two after 20 iterations and with other retailers after 50 iterations. This finding indicates that prosumers are more focused on local transactions with each other rather than with retailers. As shown in part (b) of this figure, the prices of all transactions lie above the wholesale market price. Thus, this market seems attractive for the participation of profit-centric retailers, as it enables the retailers to maximize their profits by purchasing energy from the wholesale market and selling it to local consumers. Figure 5.3 displays the number of bilateral energy transactions between all market players, with part (A) demonstrating energy trading between local sellers and buyers, part (B) between retailers and local buyers, and part (C) between the local sellers and retailers. This figure reflects the competitiveness of the proposed market.

Controlled pricing to persuade the players to engage in local trading is not applied here. Each market player trades energy with trailers or other prosumers based on its objective function. Part (A) of this figure gives the energy trading between all sellers

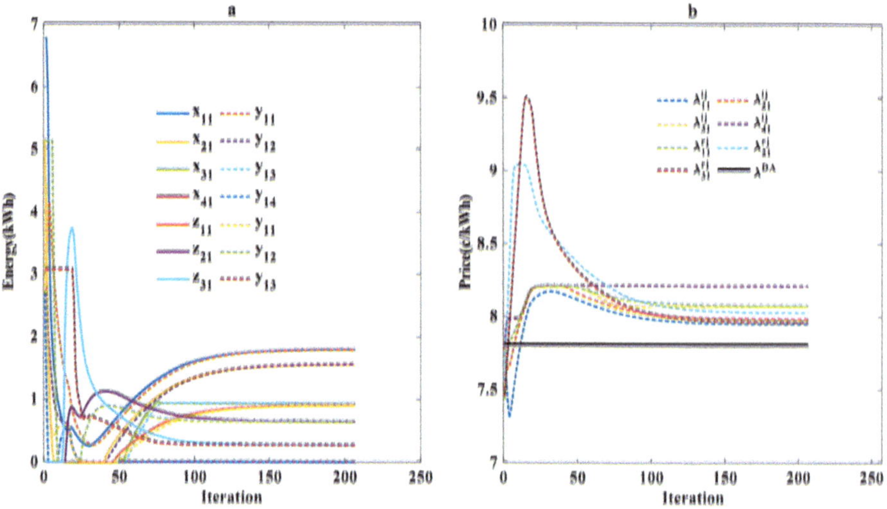


**Fig. 5.2** Evaluation of energy trading between buyer 1 and all local sellers and retailers

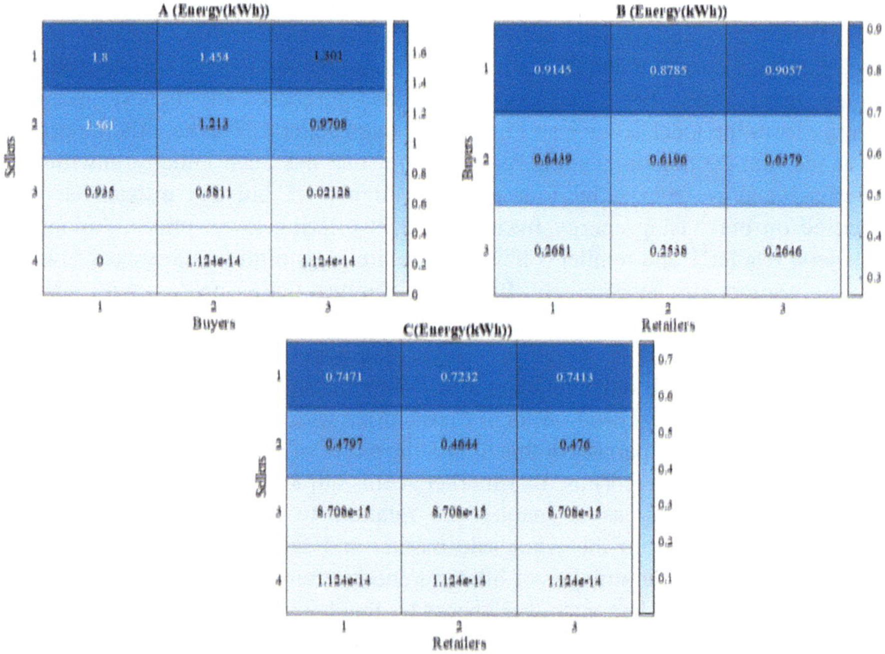


**Fig. 5.3** Bilateral energy trading between all players. (**a**) Energy trading between buyer and seller prosumers; (**b**) Energy trading between buyer prosumers, and retailers; (**c**) Energy trading between seller prosumers and retailers

**Table 5.2** The welfare of prosumers and income of retailers in the proposed P2P energy market

| Welfare (¢) | $WS_1$ | $WS_2$ | $WS_3$ | $WS_4$ | $WB_1$ | $WB_2$ | $WB_3$ | $R_1$ | $R_2$ | $R_3$ | Total (Prosumers) | Revenue |
|---|---|---|---|---|---|---|---|---|---|---|---|---|
| P2P Market | 17.23 | 7.65 | 0.06 | 0.00 | 0.65 | 16.24 | 16.29 | 0.39 | 0.38 | 0.37 | 58.11 | 1.15 |

**Table 5.3** Comparison of the results of the proposed decentralized and centralized methods

| Method | Proposed method | Centralized |
|---|---|---|
| Welfare (¢) | 59.26 | 60.05 |

and buyers. Buyer one has purchased 1.8 kWh of energy from seller one, 1.561 kWh from seller two, 0.935 kWh from seller three, and 0.0 kWh from seller four. From part B of this figure, the same buyer has purchased 0.9145 kWh of energy from retailer one, 0.8785 kWh from retailer two, and 0.9057 kWh from retailer three. Given the low decision coefficients, this player gains a higher share of the proposed market than other players.

The welfare of all proposed market players is provided in Table 5.2. The results suggest that compared to case study 1 (in the absence of retailers), the sellers' welfare increases while the buyers' welfare decreases. However, overall, the total welfare has improved. Considering the heuristic nature of the proposed approach, a final solution close to the optimal solution is expected to be found based on the stopping criterion. Table 5.3 compares the results of the proposed decentralized and centralized approaches and suggests the convergence of the proposed algorithm to the optimal solution.

### *5.3.4 Case Study 3: P2P Energy Trading with the Presence of the Retailers and Variable Wholesale Price*

This case study is similar to case study 2 in including all proposed players but evaluates the impact of the upstream market price variations on energy transactions between players and their overall welfare and income.

Table 5.4 exhibits the effect of price variations on the welfare of all prosumers and retailers. The results suggest that while isolation or imposing controlled prices may be a good incentive for local players, it will eventually impede the formation of a free competitive market and will overall be detrimental to total welfare. The proposed model assumes each prosumer can adopt the role of seller or buyer, meaning that they can act as buyers at some hours and sellers at others, based on their net generation and consumption. Figure 5.4 displays the effect of the upstream market price variations on the energy management of each studied retailer.

According to this figure, when the wholesale market price is low, the retailers purchase energy from this market and sell it to local consumers. For instance, when

**Table 5.4** The effect of upstream market price variations on income and welfare of retailers and prosumers, respectively

| | Welfare(¢) | | | | | | | | | | | |
|---|---|---|---|---|---|---|---|---|---|---|---|---|
| $\lambda^{DA}$(¢) | $WS_1$ | $WS_2$ | $WS_3$ | $WS_4$ | $WB_1$ | $WB_2$ | $WB_3$ | $R_1$ | $R_2$ | $R_3$ | Total | Revenue |
| 4 | 0.2 | 0.0 | 0.0 | 0.0 | 21.6 | 30.4 | 25.0 | 3.0 | 2.9 | 2.9 | 77.2 | 8.8 |
| 5 | 2.4 | 0.0 | 0.0 | 0.0 | 17.1 | 27.4 | 23.1 | 1.5 | 1.4 | 1.5 | 70.0 | 4.4 |
| 7 | 13.2 | 4.6 | 0.0 | 0.0 | 5.0 | 19.3 | 18.2 | 0.6 | 0.6 | 0.6 | 60.3 | 1.7 |
| 8 | 17.6 | 7.9 | 0.1 | 0.0 | 0.5 | 15.9 | 16.1 | 0.4 | 0.4 | 0.3 | 58.1 | 1.1 |
| 9 | 19.6 | 9.4 | 1.1 | 0.1 | 0.0 | 14.5 | 15.3 | 1.8 | 1.7 | 1.8 | 59.9 | 5.4 |
| 10 | 22.6 | 11.8 | 3.2 | 1.4 | 0.0 | 12.2 | 13.9 | 4.7 | 4.4 | 4.9 | 65.1 | 14.0 |
| 11 | 28.6 | 16.4 | 7.1 | 3.7 | 0.0 | 7.7 | 11.2 | 5.4 | 5.2 | 13.2 | 74.7 | 23.8 |
| 13 | 39.9 | 25.1 | 14.5 | 8.0 | 0.0 | 0.1 | 6.1 | 8.1 | 7.7 | 30.4 | 93.7 | 46.2 |
| 16 | 54.1 | 36.0 | 24.0 | 13.6 | 0.0 | 0.0 | 0.0 | 17.8 | 27.6 | 59.1 | 127.7 | 104.4 |
| 17 | 57.2 | 38.6 | 26.2 | 14.9 | 0.0 | 0.0 | 0.0 | 23.3 | 39.9 | 70.9 | 136.9 | 134.1 |
| 20 | 77.2 | 53.9 | 39.4 | 22.5 | 0.0 | 0.0 | 0.0 | 43.8 | 70.1 | 96.9 | 193.0 | 210.7 |

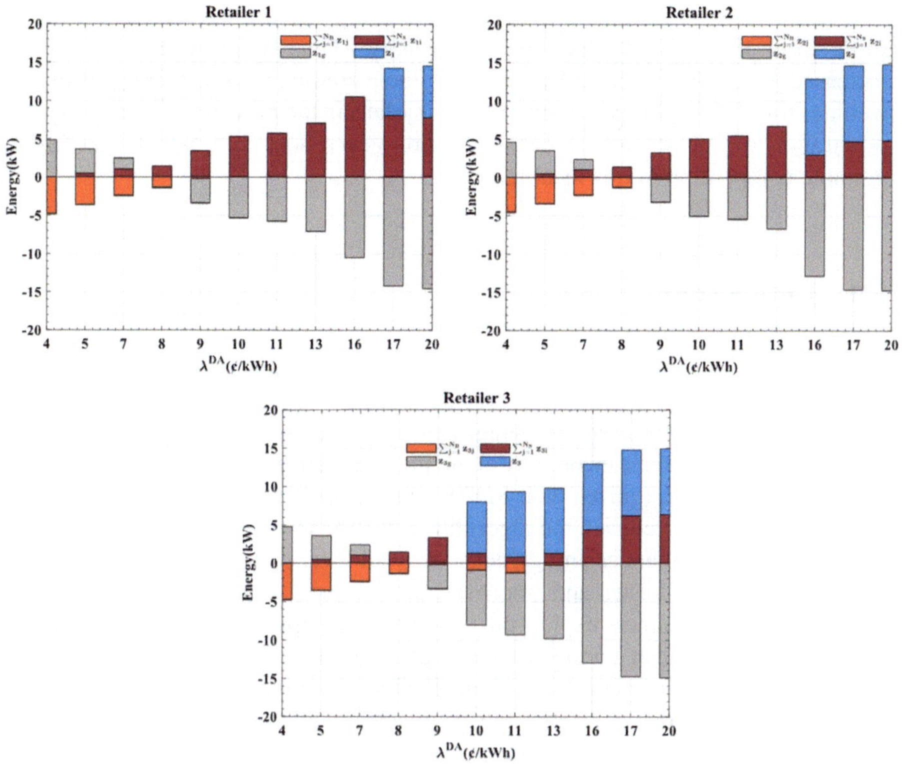


**Fig. 5.4** The effect of upstream market price variations on retailers' transactions

the wholesale market price is 4 cents, retailer one buys energy from the wholesale market and sells it to local buyers. As Table 5.4 suggests, local sellers are unwilling to participate in the market at this price, and their welfare has become zero. Conversely, all local buyers participate in the market.

At low upstream market prices, the retailers try to purchase energy from this market. For example, at 4, 5, and 7 cents prices, all retailers purchase energy from this market and sell it to local consumers. According to Table 5.4, the buyers' welfare reaches its highest value at these prices because they can gain higher utility at lower costs due to market competitiveness and the presence of retailers as energy peers. In contrast, given the high price for sellers, their participation at this price is minimal. As the upstream market price rises, the retailers participate in the local market as buyers and try to acquire higher shares from this market. Even at some prices, they self-generate and sell their excess energy to the upstream market, besides buying energy from the local market. All consumers' loads are assumed flexible in this chapter. As a result, the local buyers reduce their consumption and even do not consume energy at some prices, given the increased cost and reduced utility. For this reason, their welfare becomes zero at some prices. In contrast, an opportunity is opened for seller prosumers to increase their energy generation and gain a higher share from the offered economic opportunity. Thus, their welfare rises with increased prices. As indicated by Table 5.4, the increased price will enhance overall welfare. An outcome of the spread of these markets in the near future is the increased penetration of renewable sources with zero generation cost, which can highly promote competition in this local market.

## 5.4 Conclusion

In this chapter, a competitive market for P2P energy trading in a small distribution network in the presence of retailers was designed and cleared using a decentralized approach. The numerical studies suggested that retailers, as profit-centric entities, sought to maximize their income from the local and upstream markets. Conversely, prosumers overall maximized their total welfare by participating in the market. The presence of retailers reduced the algorithm convergence speed by over 50 percent. Interaction with retailers as energy peers allowed local prosumers more freedom to reach an agreement faster and improved their overall welfare, meaning that the retailers helped prosumers exploit the benefits of the upstream market. The numerical studies revealed that a primal-dual sub-gradient method is a reliable approach for markets with a quadratic convex mathematical model because it achieves almost the same optimal solutions as centralized approaches while preserving the privacy of decision-making data of prosumers.

## References


1. Ehsan, A., & Yang, Q. (2018). Optimal integration and planning of renewable distributed generation in the power distribution networks: A review of analytical techniques. *Applied Energy, 210*(December 2017), 44–59. https://doi.org/10.1016/j.apenergy.2017.10.106
2. Parag, Y., & Sovacool, B. K. (2016). Electricity market design for the prosumer era. *Nature Energy, 1*(4). https://doi.org/10.1038/nenergy.2016.32
3. Tushar, W., et al. (2020). A coalition formation game framework for peer-to peer energy trading. *Applied Energy, 261*(December 2019), 114436. https://doi.org/10.1016/j.apenergy.2019.114436
4. Xiao, Y., Wang, X., Pinson, P., & Wang, X. (2018). A local energy market for electricity and hydrogen. *IEEE Transactions on Power Apparatus and Systems, 33*(4), 3898–3908. https://doi.org/10.1109/TPWRS.2017.2779540
5. Xiao, Y., Xiao, Y., Wang, X., Pinson, P., & Wang, X. (2020). Transactive energy based aggregation of prosumers as a retailer. *IEEE Transactions on Smart Grid, 11*(4), 3302–3312. https://doi.org/10.1109/TSG.2020.2976130
6. Khorasany, M., Mishra, Y., & Ledwich, G. (2018). Market framework for local energy trading: A review of potential designs and market clearing approaches. *IET Generation Transmission and Distribution, 12*(22), 5899–5908. https://doi.org/10.1049/iet-gtd.2018.5309

7. Mehdinejad, M., Shayanfar, H., & Mohammadi-Ivatloo, B. (2021). Peer-to-peer decentralized energy trading framework for retailers and prosumers. *Applied Energy, 308*(April), 2022. https://doi.org/10.1016/j.apenergy.2021.118310
8. Khorasany, M., Mishra, Y., & Ledwich, G. (2020). Hybrid trading scheme for peer-to-peer energy trading in transactive energy markets. *IET Generation Transmission and Distribution, 14*(2), 245–253. https://doi.org/10.1049/iet-gtd.2019.1233
9. M. Multibuyer, S. Distribution, R. Deng, Z. Yang, and F. Hou, "Distributed real-time demand response in," pp. 1–11, 2014.
10. Khorasany, M., Mishra, Y., & Ledwich, G. (2020). A decentralized bilateral energy trading system for peer-to-peer electricity markets. *IEEE Transactions on Industrial Electronics, 67*(6), 4646–4657. https://doi.org/10.1109/TIE.2019.2931229
11. Zhu, M., & Martinez, S. (2012). On distributed convex optimization under inequality and equality constraints. *IEEE Transactions on Automatic Control, 57*(1), 151–164. https://doi.org/10.1109/TAC.2011.2167817
12. Zhang, H., Shi, W., Mokhtari, A., Ribeiro, A., & Ling, Q. (2017). Decentralized constrained consensus optimization with primal dual splitting projection. In *2016 IEEE Glob. Conf. Signal Inf. Process. Glob. 2016 - Proc* (pp. 565–569). https://doi.org/10.1109/GlobalSIP.2016.7905905
13. Morstyn, T., Mcculloch, M. D., & Member, S. Multi-class energy management for peer-to-peer energy trading driven by prosumer preferences. *IEEE Transactions on Power Systems, 99*, 1–9.
14. Sorin, E., Bobo, L., & Pinson, P. (2019). Consensus-based approach to peer-to-peer electricity markets with product differentiation. *IEEE Transactions on Power Apparatus and Systems, 34*(2), 994–1004. https://doi.org/10.1109/TPWRS.2018.2872880
15. Hug, G., Kar, S., & Wu, C. (2015). Consensus + innovations approach for distributed multiagent coordination in a microgrid. *IEEE Transactions on Smart Grid, 6*(4), 1893–1903. https://doi.org/10.1109/TSG.2015.2409053
16. Paudel, A., Sampath, L. P. M. I., Yang, J., & Gooi, H. B. (2020). Peer-to-peer energy trading in smart grid considering power losses and network fees. *IEEE Transactions on Smart Grid, 11*(6), 4727–4737. https://doi.org/10.1109/TSG.2020.2997956
17. Cui, S., Wang, Y. W., Shi, Y., & Xiao, J. W. (2020). A new and fair peer-to-peer energy sharing framework for energy buildings. *IEEE Transactions on Smart Grid, 11*(5), 3817–3826. https://doi.org/10.1109/TSG.2020.2986337
18. Aznavi, S., Fajri, P., Shadmand, M. B., & Khoshkbar-Sadigh, A. (2020). Peer-to-peer operation strategy of PV equipped office buildings and charging stations considering electric vehicle energy pricing. *IEEE Transactions on Industry Applications, 56*(5), 5848–5857. https://doi.org/10.1109/TIA.2020.2990585
19. Mehdinejad, M., Shayanfar, H. A., Mohammadi-Ivatloo, B., & Nafisi, H. (2022). Designing a robust decentralized energy transactions framework for active prosumers in peer-to-peer local electricity markets. *IEEE Access, 10*, 26743–26755. https://doi.org/10.1109/ACCESS.2022.3151922
20. Mehdinejad, M., Shayanfar, H., & Mohammadi-Ivatloo, B. (2022). Decentralized blockchain-based peer-to-peer energy-backed token trading for active prosumers. *Energy, 244*(xxxx). https://doi.org/10.1016/j.energy.2021.122713
21. Sharifi, R., Anvari-Moghaddam, A., Fathi, S. H., & Vahidinasab, V. (2020). A bi-level model for strategic bidding of a price-maker retailer with flexible demands in day-ahead electricity market. *International Journal of Electrical Power & Energy Systems, 121*(March). https://doi.org/10.1016/j.ijepes.2020.106065
22. Zhou, Y., Yu, W., Zhu, S., Yang, B., & He, J. (2020). Distributionally robust chance-constrained energy management of an integrated retailer in the multi-energy market. *Applied Energy, 286*, 116516, 2021. https://doi.org/10.1016/j.apenergy.2021.116516

23. Campos, J., & Qiao, W. (2018). A stochastic decision-making model for an electricity retailer with intermittent renewable energy and short-term demand response. *IEEE Transactions on Smart Grid, 3053*(c), 1. https://doi.org/10.1109/TSG.2018.2805326
24. Song, X., & Qu, J. (2015, March). An improved real-time pricing algorithm based on utility maximization for smart grid. In *Proc. World Congr. Intell. Control Autom* (pp. 2509–2513). https://doi.org/10.1109/WCICA.2014.7053118
25. John, J., Grainger, J., & Stevenson, W. D. (2003). *Power system analysis* (1st ed.). McGraw-Hill.
26. Peleato, B., Boyd, J. E. S., Parikh, N., & Chu, E. (2011). *Distributed optimization and statistical learning via the alternating direction method of multipliers*. Now Publishers Inc.

# Chapter 6
# The Blockchain-Based Competitive Market for Peer-to-Peer Energy Token Trading Using Demurrage Mechanism

**Mehdi Mehdinejad, Mohammad Seyfi, Mostafa Yaghoubi, Zbigniew Dziong, and Heidarali Shayanfar**

## 6.1 Introduction

### *6.1.1 Background and Motivation*

The development of electrical energy systems has made electricity the first choice of human beings for fulfilling their energy needs. This is mainly because of the feature of the high transmission speed of the electricity, which is a result of the wave nature of the electricity. Today, electrical energy has become an inevitable part of our households, and it is impossible to imagine a day without it. Because of the technological advancements, it is expected that as time goes on, the involvement of electricity in our lives is going to increase and become more effective. For example, electric vehicles are mitigating the need for other energy resources in transportation systems, like gasoline and compressed natural gas (CNG). Also, their battery capacity enables us to shift our demand from one hour to the next. The emergence of solar panels on the rooftops of residential consumers is another recent influence of electrical energy in our living environments.

These changes and the advent of distributed energy resources [1] resulted in the conversion of conventional consumers into active players in energy markets, called prosumers [2]. Prosumers are the end-consumers who have their own distributed

M. Mehdinejad (✉) · M. Seyfi · H. Shayanfar
Center of Excellence for Power Systems Automation and Operation, School of Electrical Engineering, Iran University of Science and Technology, Tehran, Iran

M. Yaghoubi · Z. Dziong
Department of Electrical Engineering, Ecole de Technologie Superieure (ETS), Montreal, QC, Canada
e-mail: mostafa.yaghoubi.1@ens.etsmtl.ca; zbigniew.dziong@etsmtl.ca

 
V. Vahidinasab, B. Mohammadi-Ivatloo (eds.), *Demand-Side Peer-to-Peer Energy Trading*, Green Energy and Technology, https://doi.org/10.1007/978-3-031-35233-1_6

energy resources. They can produce, control, moderate, and store electrical energy and act as local energy producers. As renewable energy resources are one of the main features of prosumers, they can decrease the pollutant emissions of power systems. The emergence and expansion of prosumers leads to a significant change in the power systems' structure. Managing and operating many prosumers will impose a vast calculation burden on the distribution system operator (DSO). Furthermore, prosumers are independent, profit-based users and they have their priorities to consider in the decision-making process [3]. They also seek for their privacy protection, which is neglected in the conventional central operating frameworks.

Several notions, such as microgrids, virtual power plants, energy hubs, and virtual energy hubs, have been introduced to provide the local operation of distributed energy resources. Although these approaches can facilitate the scheduling of DERs, they do not guarantee the privacy and security of prosumers and also fail to care for their preferences. These limitations brought the decentralized markets to the attention of researchers and developers [4]. In decentralized markets, and especially peer-to-peer (P2P) energy trading markets, negotiations between prosumers are done in a decentralized fashion. In other words, prosumers search for peers in the local markets to find the best option for their energy transactions depending on their preferences and financial welfare. In this kind of markets, the power of coordinators is limited, and they cannot impose a decision on the market participants. It means that prosumers have more freedom in making decisions and scheduling their demand, energy production, and energy trading in the local energy markets. This can provide different strategies for prosumers who have various priorities and goals. Most of prosumers are mainly concerned about their profit and try their best to make the most money. On the other hand, it is possible for a prosumer to seek for more clean energy or care about helping poor households by providing their demand [5]. Therefore, P2P energy trading market can incentivize prosumers to participate in power systems scheduling by providing a free and safe trading environment for them.

For developing a fully decentralized P2P energy trading market, there are some limitations that should be addressed. First, the trading structure should be fast and accessible for prosumers. It means that prosumers must access the real-time data related to market parameters and the bids/offers of other players. Secondly, the market structure should be safe and secure the privacy of prosumers and all information related to the negotiations and deals. Blockchain can address both these issues and create an accessible, fast, secure, and fully decentralized market for local small-scale prosumers [6]. Being a distributed ledger, blockchain can enhance the trust in transactions and the security of negotiations. It also provides an easy way for prosumers to access the real-time data to make the best decisions at the right moment. This technology is a result of a chain of blocks which are connected and their status depends on other blocks. Therefore, any change in the status of blocks is observable by other blocks, and there is no room for sudden and intentional changes in this structure. Furthermore, blockchain can increase the transparency of trades between peers and create an auto-paying method for prosumers using smart contracts [7]. Smart contracts are a method of trading in blockchain technology which contains all the information related to the agreement between peers. It is also programmable and can make the execution of deals automatic and can remove the supervision of any third-party center.

In this chapter, a blockchain-based energy token trading market for small-scale residential prosumers is presented, which can provide a free, competitive, safe, fast, and transparent trading environment for market participants.

### *6.1.2 Literature Review*

In this section, the related literature about P2P energy trading markets and methods for creating their required structure is discussed. These references will be discussed and reviewed in terms of market structure and market clearing methods. Then, the P2P energy trading markets which use blockchain technology will be presented and discussed.

Decentralized energy has been studied in many works with different structures and market-clearing approaches. In [8], authors have developed a decentralized energy trading market between local sellers and buyers using the primal-dual subgradient method. They considered the capacity of lines between peers and imposed the cost when the flow of the line exceeded this capacity. In the proposed trading model of this reference, peers just communicate their proposed energy price and amount of the trading energy with other market players, and there isn't any leaked information about their objective function and priorities. In these types of markets, both the seller and buyer are price makers, and they propose their desired energy price. This fact is not respected in [9], where a novel P2P energy trading mechanism for buyers and sellers considering power losses and network usage fees at the transmission level is presented. In this reference, sellers determine the current price for deals, and buyers update their demand based on this price. In other words, the buyers have been considered as the price-taker market players. The authors of this reference used the alternating direction method of multipliers (ADMM) to clear the proposed energy trading market model. The authors of [10] have proposed a bilateral energy trading in the smart grid environment to create a decentralized market. Several types of distributed energy resources like distributed generations and demand-side service have been considered in the mentioned reference. The P2P energy trading mechanism is developed in this model to form bilateral contracts.

P2P energy trading mechanism can be used to form coalitions between local prosumers to facilitate their participation in energy and ancillary markets. The new concept of the federated power plant, a virtual power plant comprised of P2P energy trading between its prosumers, is proposed in [5]. This concept enables the coalition between local prosumers to mitigate the effects of uncertainties and economic risks of participating in markets. Also, forming coalitions can make it possible for prosumers to provide ancillary service for the distribution network. The authors of [11] have used game theory to form coalitions in P2P energy trading markets between prosumers. Prosumers in this model decide on whether they use their batteries or not to participate in P2P markets. They also search for possibilities in P2P markets to join proper coalitions to create communities. The optimality of these coalitions is discussed through proper case studies.

P2P energy trading can also be used in multi-energy systems [12]. A transactive energy market framework for prosumers in multi-carrier systems has been presented in [13]. This model is created for commercial, industrial, and residential prosumers to participate in a fully decentralized market using the P2P energy trading mechanism. A new fair P2P energy trading method has been presented for multi-energy systems in [14]. This paper has created a bilateral energy trading between commercial and residential prosumers, which can trade thermal and electrical energies with their peers. Demand response programs and storage technologies have been integrated into the proposed P2P energy trading framework. For clearing the presented trading model, the authors have used a Nash-type non-cooperative game approach.

The reviewed papers have depicted a perspective of P2P energy trading applications in smart grid and multi-energy systems. However, they did not discuss the environment and structure that can enable these transactions and required communications between greedy prosumers. The blockchain technology can be utilized for this purpose.

Many works can be found in the literature discussing different aspects of the blockchain technology. The blockchain technology was first used for developing decentralized and free trading markets for cryptocurrencies [15]. Gradually, other businesses found out that this distributed ledger can be used in doing their transactions and negotiations to do their related commerce [16]. A review of the applications, challenges, and opportunities of blockchain technology has been presented in [6]. Another review of blockchain technology, its data processing, and its applications, like smart contracts and cryptography, can be found in [17]. Recently, blockchain has been used in works related to P2P energy trading markets to enable P2P communications and transaction and increase the security of trades and the privacy protection of prosumers. A real case of using blockchain in a P2P energy trading market has been settled in New York, where a pair of consumers and producers use this structure to communicate and make smart contracts to trade electrical energy [18]. The producer uses renewable energy resource and delivers its produced energy to its peer.

Recently, many works have been published about using blockchain in P2P markets [19]. In [20], blockchain has been used to develop a fully decentralized market, where the location of prosumers is considered. In this reference, prosumers are divided into consumer and producer agents in the smart grid environment. The distances between prosumers are included in the market modeling to incentivize the prosumers to trade with their neighbors. The authors of [21] proposed a smart contract structure between prosumers in the blockchain technology, which enables P2P energy trading in retail energy markets. The related case studies have experimented in the Ethereum private chain, which is one of the leaders in the blockchain technology developments. Smart contracts and blockchain structure were used in [22] to design and manage a distributed hybrid energy system in a decentralized way. The non-cooperative game approach was used in this reference to model the interactions of users with flexible loads.

**Table 6.1** Detailed comparison between this chapter and some related works

| References | Fully decentralized | Fully P2P | Blockchain | Decentralized DRP |
|---|---|---|---|---|
| [24] | ✓ | – | – | – |
| [25] | ✓ | ✓ | – | ✓ |
| [26] | ✓ | – | – | ✓ |
| [27] | ✓ | – | – | ✓ |
| [28] | ✓ | – | – | – |
| [29] | ✓ | – | – | – |
| This chapter | ✓ | ✓ | ✓ | ✓ |

Although developing decentralized markets can offer massive benefits for residential prosumers and give them more flexibility in moderating their assets, they have some disadvantages. As prosumers are independent financial-based agents, it is possible that they cooperate with each other, which can result in collusion. It can ruin the freeness and competitiveness of these markets, especially in fully decentralized markets. To address this issue demurrage mechanism can be used to prevent these collusions. In [23], the authors have proposed to use the demurrage mechanism to make a free and competitive market by forcing market players to consume or sell their energy tokens as soon as possible.

Some references are compared with this work in more detail in Table 6.1.

### 6.1.3 Study Gaps

Based on the reviewed literature, some study gaps can be found in the works written about P2P energy token trading markets. The related study gaps can be summarized as follows:

- In some of the mentioned works, the developed markets are not fully decentralized and a supervising node exists in their market modeling. Also, in some of them, both prosumers are not price makers, and some of them follow prices that are determined by others.
- The demurrage is an effective mechanism for increasing the freeness and competitiveness of energy token trading markets. However, the only reviewed reference who considered this mechanism has employed it in a semi-decentralized market. In other words, the developed market is not a fully decentralized P2P trading market.
- Among the works reviewed in the literature review section, none has considered the decentralized DR program, which can be integrated into the decentralized energy token energy trading market. In fully decentralized markets, prosumers are independent market players who decide on their own scheduling. Therefore, their participation in decentralized markets should be considered in a decentralized fashion.

### 6.1.4 Novelties and Contributions

In this chapter, a novel, fully decentralized P2P energy token trading market using the blockchain structure and smart contracts is presented. Prosumers in this model can control their decisions and DERs and freely participate in the developed market. Moreover, they can participate in the decentralized DR program in their own decision and shift some of their demand to other hours of the day. This market structure is implemented in the blockchain structure, which is a distributed ledger and can create a fully decentralized market. As mentioned, smart contracts are considered in the proposed market model, which can remove the role of third parties and the need for supervisory nodes. In this P2P energy trading market, prosumers negotiate with their peers about a potential trade, and if it is possible, they reach an agreement to trade with each other. Also, the demurrage mechanism is considered in the proposed token energy trading market to prevent prosumers from colluding with their peers, which can cause the accumulation of energy tokens at certain hours of the day. The primal-dual subgradient method is implemented to clear the developed energy token trading market in a decentralized approach. The developed market can incentivize prosumers to participate in P2P energy trading markets and provide service to power systems. In summary, the main contributions of this chapter can be stated as follows:

- A fully decentralized blockchain-based P2P energy tokens trading market, where prosumers can participate considering their priorities
- A Fully decentralized demand response program integrated in the blockchain structure
- Considering demurrage mechanism to prevent prosumers from collusion with their peers

## 6.2 Problem Formulation

It is expected that P2P energy trading markets are going to play a significant role in the development of power systems and their related markets, like energy markets and ancillary service markets. In this chapter, an energy token trading market is developed to provide the required framework for enabling a P2P trading mechanism between residential and small-scale prosumers in the distribution network level.

Prosumers in this market are divided into two groups of small-scale buyers and sellers. Sellers are referred to as prosumers whose production is more than their demand and can sell their surplus energy tokens in the P2P trading market. On the other hand, buyers are prosumers with a negative net generation and they need more energy to fulfill their hourly demand and they can buy it in the P2P energy token trading market. The developed P2P energy token trading market structure is based on the negotiations between these separate groups of prosumers. They communicate the desired energy price and amount of trading energy tokens with other players to find the best peers who are happy with that price and trading energy tokens. Besides the local peers, prosumers in this market have the option of trading with upstream markets. Based on the prices, they decide whether to trade in the local P2P energy token trading market or in upstream markets.

Smart contracts are in charge of creating connections between possible peers to negotiate and trade with each other. First, they determine an initial price for a possible deal. Then, the two peers optimize their local model to obtain the optimum value for their desired tokens to be traded based on the determined price. In the next step, the smart contract updates the price of trade by calculating the difference of the proposed energy tokens by each peer. This process continues until the two sides agree on the details of the trade.

In the developed P2P energy token trading market, the demand of buyers is flexible, and they can shift a portion of their hourly demand to other hours. It can be seen as a decentralized demand-shifting demand response program, in which prosumers change their load and shift it based on the hourly price in the local market. It can increase the flexibility of buyers for participating in markets and reduce the cost related to their energy consumption by shifting their load from expensive to cheaper hours. This is a full-decentralized demand response program, which completely respects the priorities of prosumers. The proposed market structure can be seen in Fig. 6.1.

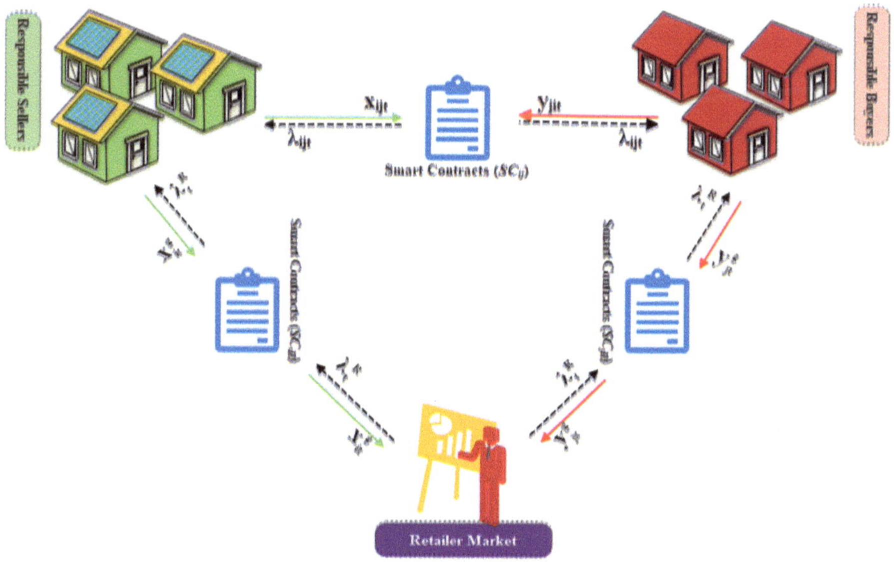


**Fig. 6.1** the developed market structure

### 6.2.1 Demurrage Mechanism

In a competitive dynamic market, the price of a product is determined by the amount of supply and demand and the abundance of that product. Since the proposed market is completely free and greedy active consumers form this market, there is a possibility of accumulation of energy tokens. In other words, buyers can get all the energy tokens they need during the hours when the price is affordable. On the other hand, energy token sellers try to keep their generated energy token and supply all of it during the hours when the energy token price is high. The demurrage mechanism considered in the transactions makes the energy token transactions very close to the hours of energy production and consumption, which prevents the accumulation of energy tokens.

If $t$ is the time of purchase of energy token by the buyers and $m$ is the time of its consumption, this mechanism can be implemented by the matrix of demurrage coefficients whose values are defined according to (6.1) and can be implemented on the transactions of the buyers of the proposed market. According to this relation, the value of energy tokens obtained from energy production by producers and energy tokens purchased by consumers decreases linearly and reaches zero after $M^{\mathrm{time}}$ hours.

If the demurrage mechanism is used with a maximum deadline of three hours for selling or consuming energy tokens, the demurrage mechanism diagram developed for this proposed market and for local buyers can be seen in Fig. 6.2. According to

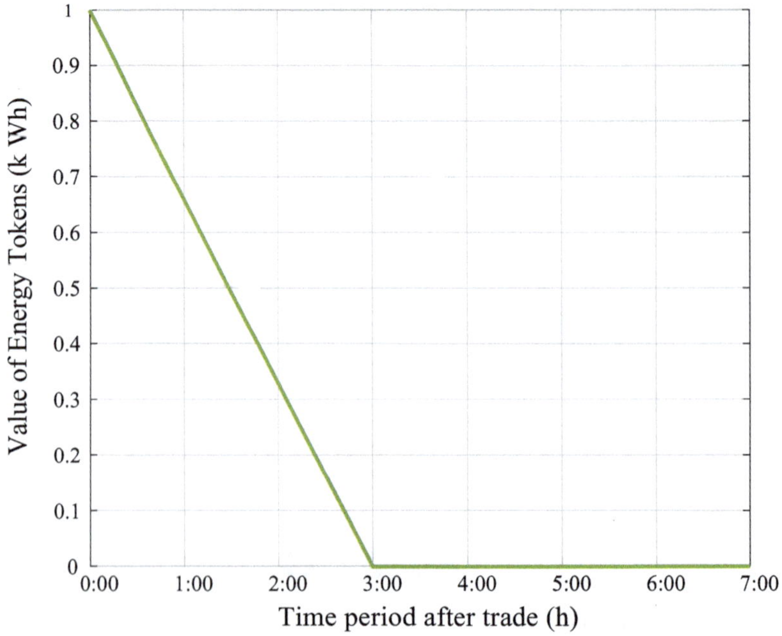


**Fig. 6.2** Values of energy tokens under the effect of demurrage mechanism

this figure, if the buyers consume the energy token in the same trading hour, the value of the energy token will not decrease. On the other hand, if there is a delay in consumption, its value will decrease linearly and after three hours it will no longer have value for consumption.

$$M_{t,m}^{\mathrm{B}} = \begin{cases} 1 - \dfrac{m-t}{M^{\mathrm{time}}} & 0 \leq m - t \leq M^{\mathrm{time}} \\ 0 & \text{otherwise} \end{cases} \tag{6.1}$$

This mechanism for sellers is obtained by transposing the matrix of buyers' demurrage coefficients according to (6.2). In this equation, $l$ is the hour of energy token production and $t$ is the hour of its sale in the peer-to-peer energy token market.

$$M_{t,l}^{\mathrm{S}} = {M_{t,m}^{\mathrm{B}}}^{\mathrm{T}} \tag{6.2}$$

## 6.2.2 *Energy Token Trading Market Model*

The developed market framework comprised of local buyers and sellers. These two groups are separate from each other and in each hour, players can be either a buyer or seller. In other words, if $\mathcal{N}_s = \{1, \ldots, N_s\}$ is the sellers set and $\mathcal{N}_b = \{1, \ldots, N_b\}$ is buyers set, then $\mathcal{N}_s \cap \mathcal{N}_b = \varnothing$. All these market players are greedy prosumers who have their own objective function and optimization problem. Modeling concepts are discussed in more detail in Chap. 5. The overall objective function of P2P energy token trading market can be obtained by summing the objective function of each market player as follows:

$$\max \left( \sum_{j=1}^{N_b} \mathrm{WB}_j + \sum_{i=1}^{N_s} \mathrm{WS}_i \right) \tag{6.3}$$

Where $\mathrm{WB}_j$ and $\mathrm{WS}_i$ are the welfare of buyers and sellers, respectively.

### 6.2.2.1 Sellers Model

Sellers are greedy prosumers, who try to reach the most profit gained from participating in the local P2P energy token trading and upstream markets. In each hour, they can sell their generated energy in these markets, based on the hourly energy price, which buyers are ready to pay. Sellers get energy tokens per each kWh energy that they produce, and then they can convert it into money by selling in the energy token trading market. Figure 6.3 shows the time period between the generation and

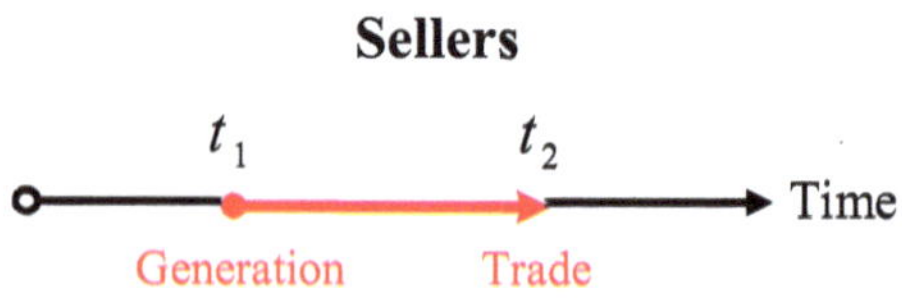


**Fig. 6.3** Time period between the generation and selling of energy tokens by sellers

sale of energy tokens of sellers. They can generate in one hour and sell it in a more expensive hour to increase their profit.

The objective function of market players should be written based on their own preferences and regardless of the social welfare of the community to model the greediness of these local sellers. The welfare of each seller can be calculated by (6.4), which consists of four terms of selling energy token to the local buyers, selling energy token to the upstream market, cost of generating energy using their DERs, and cost of using blockchain technology. $C_{i,\,t}$ which is the generation cost of sellers, can be calculated by (6.5).

$$\mathrm{WS}_i=\sum_{t=1}^{24}\left(\sum_{l=1}^{24}\sum_{j=1}^{N_b}\left(M_{t,l}^{\mathrm{S}}\times\lambda_{i,j,t}^{\mathrm{SB}}\times\mathrm{XB}_{i,j,l,t}\right)+\lambda_t^{\mathrm{UM}}\times\mathrm{XUM}_{i,t}-C_{i,t}-\pi_t\times\mathrm{XUM}_{i,t}\right)\forall i \tag{6.4}$$

$$C_{i,l}=\alpha_i\times\left(\sum_{t=1}^{24}X_{i,l,t}\right)^2+\beta_i\times\left(\sum_{t=1}^{24}X_{i,l,t}\right)+\delta_i \qquad \forall i,t \tag{6.5}$$

Where $\mathrm{XB}_{i,\,j,\,l,\,t}$, $\mathrm{XUM}_{i,\,t}$, $\pi_t$, $\lambda_t^{\mathrm{UM}}$ are energy sold by seller $i$ to buyer $j$, energy sold to upstream market, the cost rate of blockchain structure, and energy market price in the upstream market, respectively. Also, $\alpha_i$, $\beta_i$, and $\delta_i$ are parameters of sellers' cost function.

The cost of generating energy by sellers is depend on the $X_{i,\,l,\,t}$, which is the sum of the all energy generated by seller $i$ in time $l$. Equation (6.6) shows the total generation of sellers. Each seller has a limitation in its output power. In other words, their energy generation cannot exceed the maximum capacity of their DERs. Relation (6.7) demonstrates the maximum hourly power that can produced by local sellers.

$$X_{i,l,t}=\sum_{j=1}^{N_b}\mathrm{XB}_{i,j,l,t}+\mathrm{XUM}_{i,t} \qquad \forall i,t,l \tag{6.6}$$

$$\sum_{t=1}^{24}X_{i,l,t}\le X_{i,l}^{\max} \qquad \forall i,t \tag{6.7}$$

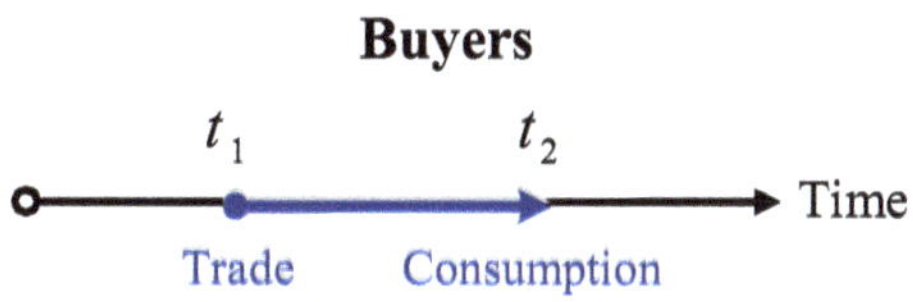


**Fig. 6.4** Time period between the buying and consumption of energy tokens by buyers

## 6.2.3 Buyers Model

Buyers in P2P token energy trading markets seek to buy their needed energy in the least price by doing negotiations with local sellers. In the developed market in this chapter, buyers have two options to buy their required energy. They can buy it from their peers in the P2P market or purchase it from upstream market. Buyers consume token energy based on the consumption of each kWh of their required demand. They can buy token energy a period before the hour that they are going to consume it. This fact is shown by Fig. 6.4.

Same as the sellers, each buyer has its unique objective function, which consists of terms that are neglecting the social welfare of all market players. The objective function of each local buyer is stated by (6.8). This objective function has four terms of cost of purchasing energy from local sellers, cost of buying energy from upstream market, the utility of consuming energy tokens, and the cost of using blockchain technology.

$$\mathrm{WB}_j = \sum_{t=1}^{24}\left(-\sum_{m=1}^{24}\sum_{i=1}^{N_b}\left(M_{t,m}^{\mathrm{B}} \times \lambda_{j,i,t}^{\mathrm{BS}} \times \mathrm{YS}_{j,i,t,m}\right) - \lambda_t^{\mathrm{UM}} \times \mathrm{YUM}_{i,t} + U_{j,t} - \pi_t \times \mathrm{YUM}_{j,t}\right)\forall j \tag{6.8}$$

Where $\mathrm{YS}_{j,\ i,\ t,\ m}$, $\mathrm{YUM}_{j,\ t}$ are energy bought by buyer $i$ to seller $j$, and energy bought from the upstream market, respectively.

For writing the objective function of buyers, it is needed to model their energy consumption behavior using a proper function, called utility function. This function should have a positive first-order derivative and a negative second-order derivative. It should satisfy constraints (6.9), (6.10), and (6.11). One utility function that follows this constraints is introduced in [30]. In this chapter, this utility function is used to model the consumption behavior of local small-scale buyers, which is given by (6.12). The total token energy consumed by buyers can be calculated using (6.13).

$$\frac{dU_{j,t}}{dy_{j,t}} \geq 0 \tag{6.9}$$

$$\frac{d^2U_{j,t}}{dy_{j,t}{}^2} \leq 0 \tag{6.10}$$

$$U_{j,t}(0) = 0 \tag{6.11}$$

$$U_{j,m} = \begin{cases} \omega_j Y_{j,m} - \theta_j {Y_{j,m}}^2, & Y_{j,m} < \dfrac{\omega_j}{2\theta_j} \\ \dfrac{{\omega_j}^2}{2\theta_j}, & Y_{j,m} \geq \dfrac{\omega_j}{2\theta_j} \end{cases} \quad \forall j, m \tag{6.12}$$

$$\text{Load}_{j,m} + Y^{\text{DR}}_{j,t} = \sum_{i=1}^{N_s} \sum_{t=1}^{24} M^{\text{B}}_{t,m} \times \text{YS}_{j,i,t,m} + \text{YUM}_{j,t} \quad \forall j, m \tag{6.13}$$

Where, $\omega_j$ and $\theta_j$ are parameters of buyers' utility function. Also, $Y_{j,\ m}$ is the total energy token bought by buyers, $Y^{\text{DR}}_{j,t}$ is the shifted load of buyers, and $\text{Load}_{j,\ m}$ is buyers' forecasted load, respectively.

As said before, buyers can participate in the demand response program, and move some part of their demand to other hours. Constraints (6.14) and (6.15) state the mathematical modeling of proposed demand response program.

$$-\text{DR}_{\text{max}} \times \text{Load}_{j,m} \leq Y^{\text{DR}}_{j,t} \leq \text{DR}_{\text{max}} \times \text{Load}_{j,m} \quad \forall j, t \tag{6.14}$$

$$\sum_{t=1}^{N_t} Y^{\text{DR}}_{j,t} = 0 \quad \forall j \tag{6.15}$$

Where $\text{DR}_{\text{max}}$ is the maximum percentage of hourly load considered for the demand response program.

## 6.3 Market Clearing Process

To clear the developed fully decentralized P2P token energy trading market, it is necessary to utilize a distributed optimization method to solve the local models of prosumers separately. In this chapter, the primal-dual subgradient optimization method [31] is used to clear the transactions in the market and optimize the model of prosumers. The main central optimization problem can be written as:

$$\max \left( \sum_{j=1}^{N_b} \text{WB}_j + \sum_{i=1}^{N_s} \text{WS}_i \right) \tag{6.16}$$

Subject to

$$\sum_{l=1}^{N_t} M^{\text{S}}_{t,l} \times \text{XB}_{i,j,l,t} = \sum_{m=1}^{N_t} \text{YS}_{j,i,t,m} \tag{6.17}$$

Equations (6.5), (6.6), (6.7) and (6.12), (6.13), (6.14), (6.15).

Equation (6.17) is the coupling constraint of the main problem, which ensures the equality of the proposed trading energy tokens by sellers and buyers.

This model is a central optimization problem, which neglects the privacy and preferences of prosumers. As stated before, this chapter aims to develop a fully decentralized P2P energy token trading market for active prosumers. Therefore, the central problem should be divided into the subproblems of sellers and buyers. We use primal-dual subgradient method to decentralize the market model. The primal-dual subgradient model method uses lagrangian of the optimization problem to relax the constraints. The lagrangian of the optimization problem with objective function stated in (6.3) can be written as:

$$
\begin{aligned}
&\ell\left(\mathrm{XB}_{i,j,l,t}, \mathrm{YS}_{j,i,t,m}, \lambda_{i,j,t}, \lambda_t^{\mathrm{UM}}, \mathrm{YUM}_{j,t}, Y_{j,t}^{\mathrm{DR}}, \sigma_{jt}, \vartheta_{jt}, \overline{\mu}_{jt}, \underline{\mu}_{jt}, \overline{\mu}_{it}, \underline{\mu}_{it}\right) \\
&= \sum_{t=1}^{N_t}\left(\sum_{j=1}^{N_b}\left(U_{j,t} - \lambda_t^{\mathrm{UM}} \times \mathrm{YUM}_{i,t} - \pi_t \times \mathrm{YUM}_{j,t}\right)\right. \\
&\quad + \sum_{i=1}^{N_s}\left(\lambda_t^{\mathrm{UM}} \times \mathrm{XUM}_{i,t} - C_{i,t} - \pi_t \times \mathrm{XUM}_{i,t}\right) \\
&\quad + \sum_{j=1}^{N_b}\sum_{i=1}^{N_s}\left(\lambda_{i,j,t} \times \left(\sum_{l=1}^{N_t} M_{t,l}^{\mathrm{S}} \times \mathrm{XB}_{i,j,l,t} - \sum_{m=1}^{N_t} \mathrm{YS}_{j,i,t,m}\right)\right) \\
&\quad - \sum_{j=1}^{N_{\mathrm{B}}}\sum_{t=1}^{T} \theta_{jt}\left(\sum_{i=1}^{\mathcal{N}_{\mathrm{S}}}\sum_{\tau=1}^{T} M_{t,m}^{\mathrm{B}} \mathrm{YS}_{j,i,t,m} + \mathrm{YUM}_{j,t} - \left(\mathrm{Load}_{j,m} + Y_{j,t}^{\mathrm{DR}}\right)\right) \\
&\quad - \sum_{j=1}^{N_{\mathrm{B}}} \vartheta_j\left(\sum_{t=1}^{T} Y_{j,t}^{\mathrm{DR}}\right) - \sum_{j=1}^{N_{\mathrm{B}}}\sum_{t=1}^{T} \overline{\mu}_{jt}\left(Y_{j,t}^{\mathrm{DR}} - \mathrm{DR}_{\max} \times \mathrm{Load}_{j,m}\right) \\
&\quad - \sum_{j=1}^{N_{\mathrm{B}}}\sum_{t=1}^{T} \underline{\mu}_{jt}\left(-\mathrm{DR}_{\max} \times \mathrm{Load}_{j,m} - y_{jt}^{\mathrm{DR}}\right) - \sum_{i=1}^{N_{\mathrm{S}}}\sum_{t=1}^{T} \overline{\mu}_{it}\left(\mathrm{XB}_{i,j,l,t} - X_{i,l}^{\max}\right) \\
&\quad \left. - \sum_{i=1}^{N_{\mathrm{S}}}\sum_{t=1}^{T} \underline{\mu}_{it}\left(X_{i,l}^{\min} - \mathrm{XB}_{i,j,l,t}\right)\right)
\end{aligned}
\tag{6.18}
$$

In this equation, $\lambda_{i,\,j,\,t}$ plays as the dual variable of the dual optimization problem, and its value is related to the difference of the energy tokens bought and sold by buyers and sellers, respectively.

In the primal dual subgradient method, iteration-based relations are used to calculate the dual variables value based on the bids and offers of buyers and sellers and their optimization problem. The dual variable value is calculated by taking derivative of mentioned objective function with respect to each one as follows:

$$\lambda_{i,j,t}{}^{k+1} = \lambda_{i,j,t}{}^{k+1} - \rho\left(\sum_{l=1}^{N_t} M_{t,l}^{\mathrm{S}} \times \mathrm{XB}_{i,j,l,t} - \sum_{m=1}^{N_t} \mathrm{YS}_{j,i,t,m}\right) \tag{6.19}$$

$$\sigma_{jt}^{k+1} = \left[\sigma_{jt}^{k} + \Gamma\left(\sum_{i=1}^{\mathcal{N}_S} \sum_{\tau=1}^{T} M_{t,m}^{\mathrm{B}} \times \mathrm{YS}_{j,i,t,m} + \mathrm{YUM}_{j,t} - \left(Y_{j,m} + Y_{j,t}^{\mathrm{DR}}\right)\right)\right]^{\pm} \tag{6.20}$$

$$\vartheta_{jt}^{k+1} = \left[\vartheta_{jt}^{k} + \Gamma\left(\sum_{t=1}^{T} Y_{j,t}^{\mathrm{DR}}\right)\right]^{\pm} \tag{6.21}$$

$$\bar{\mu}_{it}^{k+1} = \left[\bar{\mu}_{it}^{k} + \phi\left(\mathrm{XB}_{i,j,l,t} - X_{i,l}^{\max}\right)\right]^{+} \tag{6.22}$$

$$\underline{\mu}_{it}^{k+1} = \left[\underline{\mu}_{it}^{k} + \phi\left(X_{i,l}^{\min} - \mathrm{XB}_{i,j,l,t}\right)\right]^{+} \tag{6.23}$$

$$\bar{\varphi}_{jt}^{k+1} = \left[\bar{\varphi}_{jt}^{k} + \phi\left(Y_{j,t}^{\mathrm{DR}} - \mathrm{DR}_{\max} \times \mathrm{Load}_{j,m}\right)\right]^{+} \tag{6.24}$$

$$\underline{\varphi}_{jt}^{k+1} = \left[\underline{\varphi}_{jt}^{k} + \phi\left(-\mathrm{DR}_{\max} \times \mathrm{Load}_{j,m} - y_{jt}^{\mathrm{DR}}\right)\right]^{+} \tag{6.25}$$

In these equations, $\lambda_{i,\,j,\,t}{}^{k}$, $\sigma_{jt}^{k}$, $\vartheta_{jt}^{k}$, $\bar{\mu}_{it}^{k}$, $\bar{\mu}_{jt}^{k}$, $\bar{\varphi}_{jt}^{k}$, and $\underline{\varphi}_{jt}^{k}$ are the dual variables related to each equation of primal problem. Also, $\rho$, $\Gamma$, and $\phi$ are the step sizes of the optimization process, which can control the convergence speed of reaching agreements about possible deals. After updating the price value, each player solves its local model and communicates their optimum values to the smart contract to find the next updated price.

The local problem of the buyers can be stated as:

$$\max_{Y_{j,m}} \mathrm{WB}_j = \sum_{t=1}^{24}\left(-\sum_{m=1}^{24}\sum_{i=1}^{N_b}\left(M_{t,m}^{\mathrm{B}} \times \lambda_{j,i,t}^{\mathrm{BS}} \times \mathrm{YS}_{j,i,t,m}\right) - \lambda_t^{\mathrm{UM}} \times \mathrm{YUM}_{i,t}\right. \\ \left. + U_{j,t} - \pi_t \times \mathrm{YUM}_{j,t}\right) \forall j \tag{6.26}$$

Subject to

$$(6.12) - (6.15) \tag{6.27}$$

And the local problem of the sellers can be stated as:

$$\max_{X_{i,l,t}} \mathrm{WS}_i = \sum_{t=1}^{24} \times \left( \sum_{l=1}^{24} \sum_{j=1}^{N_b} \left( M_{t,l}^{\mathrm{S}} \times \lambda_{i,j,t}^{\mathrm{SB}} \times \mathrm{XB}_{i,j,l,t} \right) + \lambda_t^{\mathrm{UM}} \times \mathrm{XUM}_{i,t} - C_{i,t} - \pi_t \times \mathrm{XUM}_{i,t} \right) \quad \forall i \tag{6.28}$$

Subject to

$$(6.5) - (6.7) \tag{6.29}$$

## 6.4 Simulation Results

In this section, the proposed P2P energy token trading market model is simulated and the results will be discussed. The GAMS software is used to optimize the local optimization problems of prosumers and create the iteration-based approach. This model is simulated using a system with an Intel Core i7 processor and 8 gigabytes of RAM.

The developed and modeled market structure will be studied in one case study. In this case study, the market is cleared using a fixed price in the upstream market. The market is cleared for 24 hours, with a time resolution of 1 hour. The different aspects of this market, such as average energy price, energy transactions between local prosumers in the local market and upstream market, demand response program, and the effects of demurrage mechanism is discussed through this case study.

### *6.4.1 Input Data*

In this chapter, a small-scale market model consisting of four local sellers and three local buyers is modeled and simulated. Sellers are equipped with solar panels, and they can sell energy in the middle hours of the day. Buyers try to buy their required energy by participating in the P2P market. The data related to this prosumers is obtained from [32]. The maximum output power of sellers, buyers load profile, and blockchain usage costs are shown in Fig. 6.5. Another parameter which is crucial in the transactions of local P2P market is the upstream market price. In this chapter, the average of hourly prices in the New York market is used as a fixed price. The price of using blockchain structure depends on the amount of local generation to encourage local prosumers to trade in the local market instead of upstream markets when it is possible to make deals in this market.

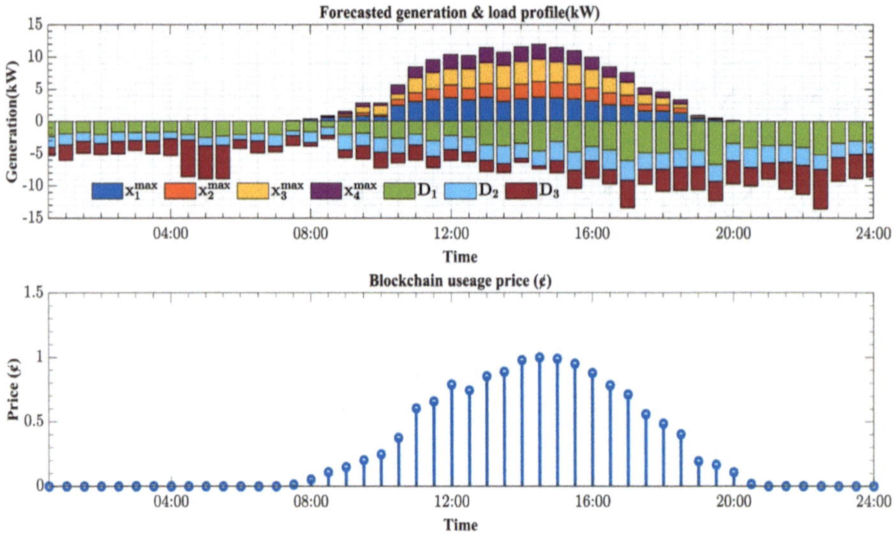


**Fig. 6.5** buyers demand, sellers energy capacity, and the blockchain usage price

## 6.4.2 *Case Study*

In this case study, the P2P energy token trading mechanism between small-scale residential prosumers is simulated using the fixed price in the upstream market during the 24 hours of the day. In this situation, prosumers always have a steady measure to decide whether to participate in the P2P energy trading market or in the local market.

In this situation, transactions between market players are shown in Fig. 6.6. The focus of this figure is on transactions of sellers. To show the convergence speed of market clearing stage, the results have been drawn for four different time slots. It can be seen that all transactions have been converged after about 5000 iterations, which is the closing of the negotiations by reaching an agreement. The sub-plot related to hour 10 o'clock shows all transactions for seller #1 with local buyers. It can be seen that this seller has decided to sell a percentage of its production to different buyers. Therefore, any market player can have different peers and reach more than an agreement with other players. They choose different strategies in negotiations with various market players. This case has been repeated for other sellers too. For example, the sub-plot of time slot 15:30 shows the transactions of seller #3, which have done multiple trades in the local market. Overall, this figure proves the convergence of proposed market clearing approach using primal-dual subgradient method. The advantage of this method is in the fact that any player has only optimized its own optimization model with its unique objective function.

One state which is notable is the less numbers of iterations needed for reaching agreements in the sub-plot of time slot 17:30. The reason for this is the demand of

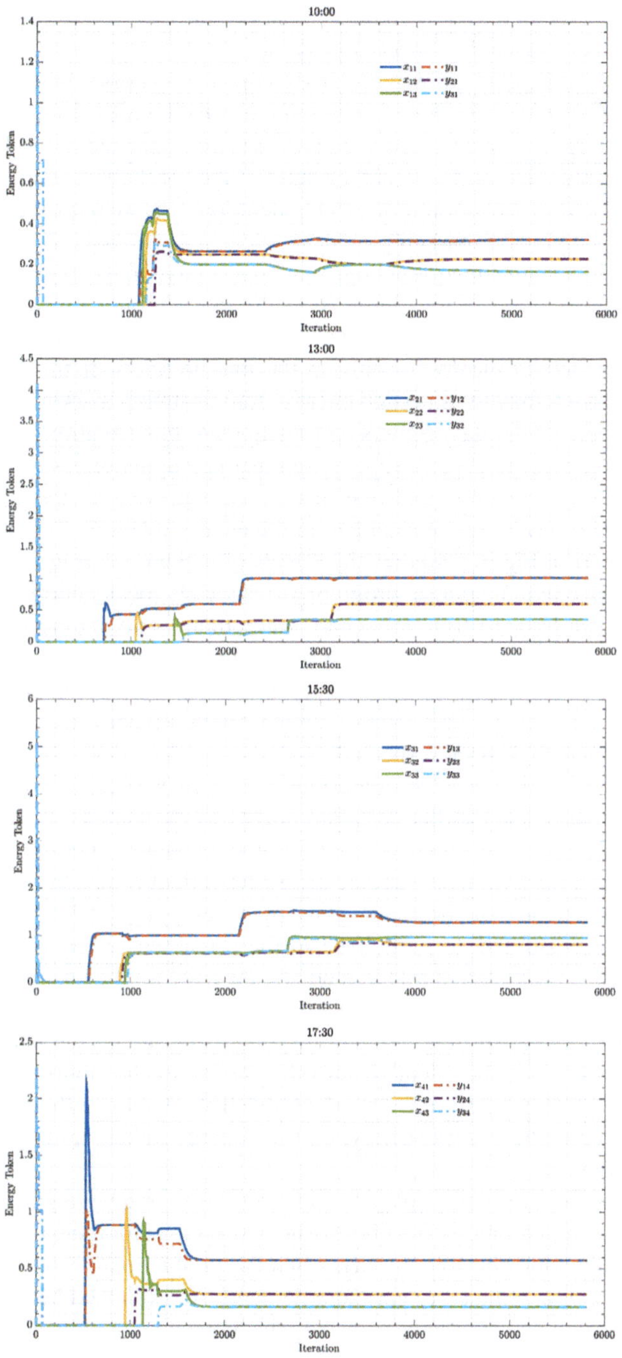


**Fig. 6.6** Energy token transactions in the local P2P market in case study 1

local buyers that is increased in the later hours of the day. Therefore, they try their best to fulfill their demand as soon as possible.

The welfare of buyers is directly related to the average price in upstream market. When the market price is high sellers see the upstream market as a good option for selling their energy. Therefore, it can be harder to convince them to sell their energy in the local trading market. However, because of the blockchain usage cost which is considered related to the amount of local generation, it is more profitable for sellers to participate in the local markets.

Figure 6.7 illustrates the transactions that buyers have in the energy token trading market. First three charts in this figure show the transactions of buyer #1, buyer #2, and buyer #3 with other market players, respectively. As aforementioned, the local generation can deliver electricity only in the middle hours of the day. Therefore, local buyers have to buy their needed energy from upstream market in the beginning and ending hours of the day. Sellers can manipulate this situation by moving their energy tokens to these hours to be able to sell them in higher prices. This can ruin the dynamic nature of economic and free markets. For preventing this issue, we have implemented the demurrage mechanism in the energy token trading market.

The last chart in Fig. 6.7 studies the effects of using demurrage mechanism and load-shifting demand response program in the P2P local energy token trading market. As it can be seen from this chart, when there is no demurrage mechanism, the sellers have no transactions with upstream markets. It is because buyers have bought their required energy tokens during these hours and consumed them during the more expensive hours. This means that the competitiveness and dynamic behavior of markets have not been respected.

On the other hand, when the demurrage mechanism is considered, sellers have sold some of their energy in the upstream energy token markets. It can be seen that the generation capacity of sellers is more than the demanded energy by local buyers. Thus, it is reasonable that sellers should sell some of their production in the upstream market. However, one of the main goals of developing local P2P markets is to incentivize the prosumers to participate in these markets. Here, demand response program can play a significant role in reducing the need for the upstream markets. As shown in this chart, it can be noted that when the 20 percent of maximum demand is considered as the load-shifting demand response program on the behavior of local buyers, they can shift their load from other hours to the hours that there is an abundance of local generation by residential sellers to reduce their purchase cost, and also increase the independency of P2P energy token trading market from upstream markets.

The participation of market players can alter the average energy prices of transactions. They do negotiations and choose the best options for them to maximize their welfare. In a dynamic and free market, when the production of goods exceeds their demand, the price is reduced to get more demand. As explained before, this normal behavior of market players can be ruined by the happening of collusion among them. It can be seen in (a) sub-plot of Fig. 6.8 that when the demurrage mechanism is not considered, the price profile has a relatively steady value during all hours of the day. Even when the local generation enters the market no big change in the price variation

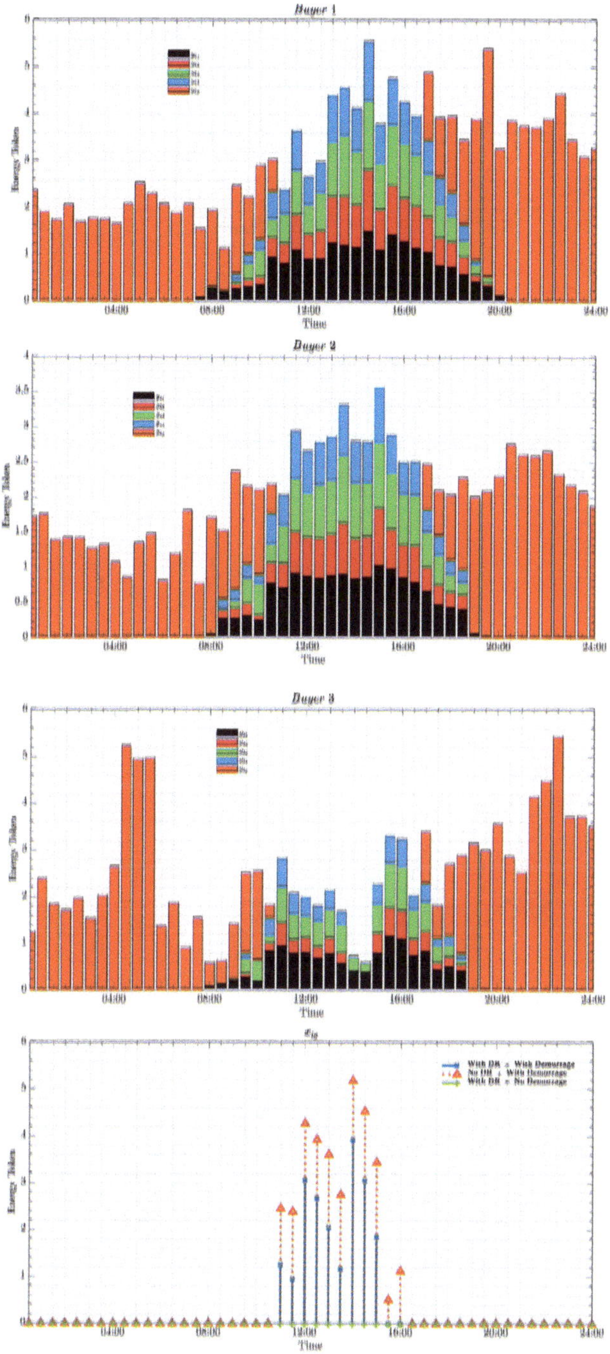


**Fig. 6.7** transactions of buyers in case study 1

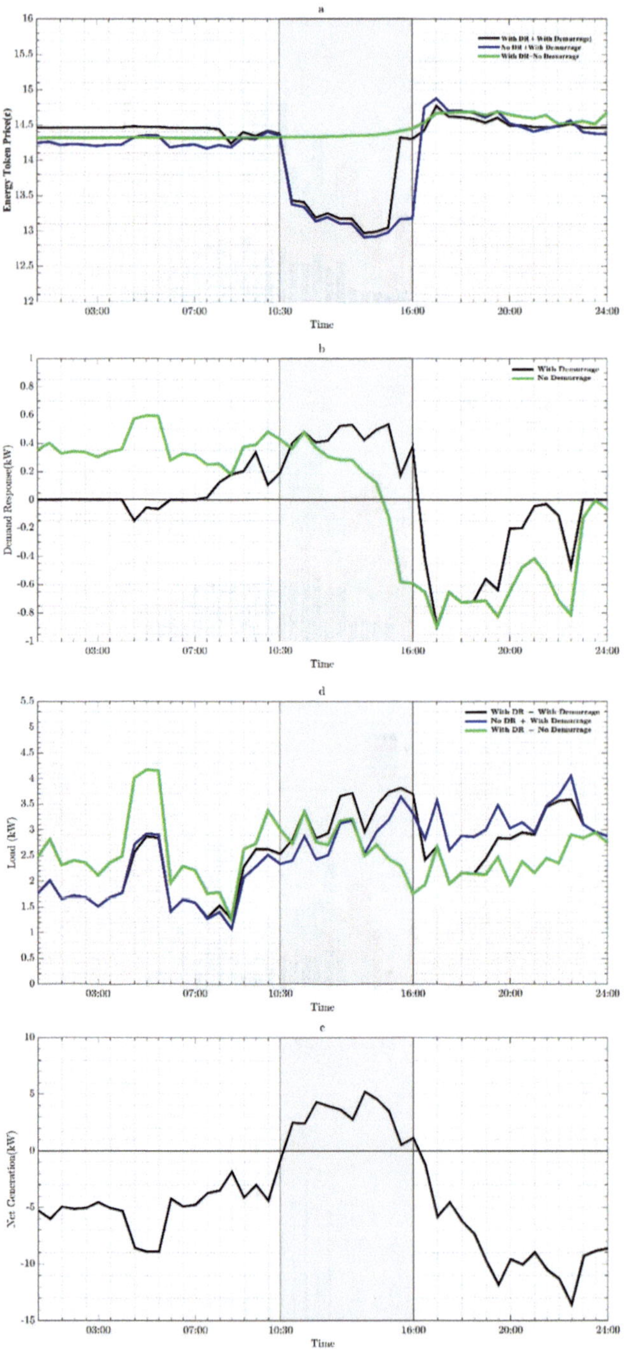


**Fig. 6.8** Effects of demurrage and demand response program

can be observed. But, when the demurrage mechanism is considered, it can be seen that a drop occurs in the price profile during the middle hours of the day, when the local generation can provide the buyers' demand. When the demand response program is enabled in the market, this drop decreases a little because buyers shift some portion of their load to this hour.

The (b) sub-plot of Fig. 6.8 shows the effect of demurrage mechanism in the performance of demand response program of buyers. It can be seen that even demurrage mechanism can enhance the demand response program result, as more energy demand has been shifted during the 24 hours of the day. This state has been shown better in the (c) sub-plot of Fig. 6.8.

The (d) sub-plot of Fig. 6.8 shows the difference between local generation and demand during the 24 hours of the day. As mentioned before, the local generation is a result of rooftop PVs that prosumers can use them to gain money. Therefore, in the middle hours of the day, we can observe a positive value for this difference, where the drop of price has occurred.

## 6.5 Conclusion

In this chapter, a blockchain-based P2P energy token trading market was developed to enable the fully decentralized transactions between small-scale prosumers. In this market, prosumers chose their peers based on their strategies. The buyers or consumers of this market have flexible demands, which provide enough flexibility for forming the demand response program. The load-shifting demand response program was considered in this structure. Also, we implemented the demurrage mechanism on the P2P energy token trading market to enhance the competitiveness of this market. The developed market was simulated using GAMS software to optimize the developed distributed optimization problem. The results showed the importance of demand response program in reducing the needs for upstream markets and effects of demurrage mechanism in preventing the token accumulation.

## References


1. Nikam, V., & Kalkhambkar, V. (2021). A review on control strategies for microgrids with distributed energy resources, energy storage systems, and electric vehicles. *International Transactions on Electrical Energy Systems*. https://doi.org/10.1002/2050-7038.12607
2. Ritzer, G., Dean, P., & Jurgenson, N. (2012). The coming of age of the prosumer. *The American Behavioral Scientist*. https://doi.org/10.1177/0002764211429368
3. Seyfi, M., Mehdinejad, M., Mohammad-Ivatloo, B., & Shayanfar, H. (2021). Decentralized peer-to-peer energy trading for prosumers considering demand response program. In *2021 11th smart grid conference SGC 2021*. https://doi.org/10.1109/SGC54087.2021.9664052
4. Zhang, C., Wu, J., Long, C., & Cheng, M. (2017). Review of existing peer-to-peer energy trading projects. *Energy Procedia*. https://doi.org/10.1016/j.egypro.2017.03.737

5. Morstyn, T., Farrell, N., Darby, S. J., & McCulloch, M. D. (2018). Using peer-to-peer energy-trading platforms to incentivize prosumers to form federated power plants. *Nature Energy.* https://doi.org/10.1038/s41560-017-0075-y
6. Monrat, A. A., Schelén, O., & Andersson, K. (2019). A survey of blockchain from the perspectives of applications, challenges, and opportunities. *IEEE Access.* https://doi.org/10.1109/ACCESS.2019.2936094
7. Christidis, K., & Devetsikiotis, M. (2016). Blockchains and smart contracts for the Internet of Things. *IEEE Access.* https://doi.org/10.1109/ACCESS.2016.2566339
8. Khorasany, M., Mishra, Y., & Ledwich, G. (2020). A decentralized bilateral energy trading system for peer-to-peer electricity markets. *IEEE Transactions on Industrial Electronics.* https://doi.org/10.1109/TIE.2019.2931229
9. Paudel, A., Sampath, L. P. M. I., Yang, J., & Gooi, H. B. (2020). Peer-to-peer energy trading in smart grid considering power losses and network fees. *IEEE Transactions on Smart Grid.* https://doi.org/10.1109/TSG.2020.2997956
10. Morstyn, T., Teytelboym, A., & McCulloch, M. D. (2019). Bilateral contract networks for peer-to-peer energy trading. *IEEE Transactions on Smart Grid.* https://doi.org/10.1109/TSG.2017.2786668
11. Tushar, W., et al. (2020). A coalition formation game framework for peer-to-peer energy trading. *Applied Energy.* https://doi.org/10.1016/j.apenergy.2019.114436
12. Seyfi, M., Mehdinejad, M., Mohammadi-Ivatloo, B., & Shayanfar, H. (2022). Deep learning-based scheduling of virtual energy hubs with plug-in hybrid compressed natural gas-electric vehicles. *Applied Energy, 321*, 119318. https://doi.org/10.1016/j.apenergy.2022.119318
13. Javadi, M. S., Esmaeel Nezhad, A., Jordehi, A. R., Gough, M., Santos, S. F., & Catalão, J. P. S. (2022). Transactive energy framework in multi-carrier energy hubs: A fully decentralized model. *Energy.* https://doi.org/10.1016/j.energy.2021.121717
14. Jing, R., Xie, M. N., Wang, F. X., & Chen, L. X. (2020). Fair P2P energy trading between residential and commercial multi-energy systems enabling integrated demand-side management. *Applied Energy.* https://doi.org/10.1016/j.apenergy.2020.114551
15. Tschorsch, F., & Scheuermann, B. (2016). Bitcoin and beyond: A technical survey on decentralized digital currencies. *IEEE Communication Surveys and Tutorials.* https://doi.org/10.1109/COMST.2016.2535718
16. Underwood, S. (2016). Blockchain beyond bitcoin. *Communications of the ACM.* https://doi.org/10.1145/2994581
17. Dinh, T. T. A., Liu, R., Zhang, M., Chen, G., Ooi, B. C., & Wang, J. (2018). Untangling Blockchain: A data processing view of blockchain systems. *IEEE Transactions on Knowledge and Data Engineering.* https://doi.org/10.1109/TKDE.2017.2781227
18. Mengelkamp, E., Gärttner, J., Rock, K., Kessler, S., Orsini, L., & Weinhardt, C. (2018). Designing microgrid energy markets: A case study: The Brooklyn microgrid. *Applied Energy.* https://doi.org/10.1016/j.apenergy.2017.06.054
19. Soto, E. A., Bosman, L. B., Wollega, E., & Leon-Salas, W. D. (2021). Peer-to-peer energy trading: A review of the literature. *Applied Energy.* https://doi.org/10.1016/j.apenergy.2020.116268
20. Khorasany, M., Dorri, A., Razzaghi, R., & Jurdak, R. (2021). Lightweight blockchain framework for location-aware peer-to-peer energy trading. *International Journal of Electrical Power & Energy Systems.* https://doi.org/10.1016/j.ijepes.2020.106610
21. Han, D., Zhang, C., Ping, J., & Yan, Z. (2020). Smart contract architecture for decentralized energy trading and management based on blockchains. *Energy.* https://doi.org/10.1016/j.energy.2020.117417
22. Li, Y., Yang, W., He, P., Chen, C., & Wang, X. (2019). Design and management of a distributed hybrid energy system through smart contract and blockchain. *Applied Energy.* https://doi.org/10.1016/j.apenergy.2019.04.132
23. Devine, M. T., & Cuffe, P. (2019). Blockchain electricity trading under demurrage. *IEEE Transactions on Smart Grid.* https://doi.org/10.1109/TSG.2019.2892554

24. Xiao, Y., Xiao, Y., Wang, X., Pinson, P., & Wang, X. (2020). Transactive energy based aggregation of prosumers as a retailer. *IEEE Transactions on Smart Grid, 11*(4), 3302–3312. https://doi.org/10.1109/TSG.2020.2976130
25. Khorasany, M., Mishra, Y., & Ledwich, G. (2018). Market framework for local energy trading: A review of potential designs and market clearing approaches. *IET Generation, Transmission and Distribution*. https://doi.org/10.1049/iet-gtd.2018.5309
26. Nojavan, S., Zare, K., & Mohammadi-Ivatloo, B. (2017). Optimal stochastic energy management of retailer based on selling price determination under smart grid environment in the presence of demand response program. *Applied Energy, 187*, 449–464. https://doi.org/10.1016/j.apenergy.2016.11.024
27. Aghamohammadloo, H., et al. (2021). Integrated Demand Response programs and energy hubs retail energy market modelling. *Energy*. https://doi.org/10.1016/j.energy.2021.121239
28. Moret, F., & Pinson, P. (2019). Energy collectives: A community and fairness based approach to future electricity markets. *IEEE Transactions on Power Apparatus and Systems, 34*(5), 3994–4004. https://doi.org/10.1109/TPWRS.2018.2808961
29. Sharifi, R., Anvari-Moghaddam, A., Fathi, S. H., & Vahidinasab, V. (2020). A bi-level model for strategic bidding of a price-maker retailer with flexible demands in day-ahead electricity market. *International Journal of Electrical Power & Energy Systems*. https://doi.org/10.1016/j.ijepes.2020.106065
30. Samadi, P., Mohsenian-Rad, H., Schober, R., & Wong, V. W. S. (2012). Advanced demand side management for the future smart grid using mechanism design. *IEEE Transactions on Smart Grid*. https://doi.org/10.1109/TSG.2012.2203341
31. Mehdinejad, M., Shayanfar, H., & Mohammadi-Ivatloo, B. (2022). Peer-to-peer decentralized energy trading framework for retailers and prosumers. *Applied Energy, 308*, 118310. https://doi.org/10.1016/J.APENERGY.2021.118310
32. Dataport | Login | Signup. https://dataport.pecanstreet.org/. Accessed 23 Dec 2022.

# Chapter 7
# P2P Energy Trading in a Community of Individual Consumers with the Presence of Central Shared Battery Energy Storage

**Ali Aminlou, Mohammad Mohsen Hayati, and Kazem Zare**

## 7.1 Introduction

The energy system in the world is moving towards decarbonization by developing more and more renewable energies [1]. Therefore, we will have a huge revolution ahead. Because the transition to a modern, low-carbon, and energy-efficient economy is happening fast [2]. During recent years, the penetration and expansion of new technologies such as distributed energy resources (DER) as a source for individual customers in conventional power systems, along with the rapid development of energy storage systems (ESS), have caused fundamental changes in energy systems and great challenges to the real-time balance between the demand and supply of electricity [3–6]. Distributed energy resources (DER), such as wind turbines, solar panels/photovoltaics (PV), and energy storage devices and electric vehicles (EVs), have been widely developed in modern power systems [7, 8]. The widespread installation of DERs in power grids provides benefits to owners and the grid. On the other hand, DER owners can also manage their energy consumption and receive revenue by selling excess energy [9]. According to several studies, the widespread penetration of renewable energy sources (RES) will exceed 60% by 2050 [10]. A decentralized peer-to-peer (P2P) energy trading mechanism is proposed to promote the interaction between distributed energy resources (DER) and interconnected microgrids (MGs) [11, 12].

From the point of view of the power grid, MG is an island microsystem that provides energy services in addition to supplying local consumers by interacting with the electricity market. On the other hand, energy storage systems (ESS), such as

A. Aminlou (✉) · M. M. Hayati · K. Zare
Faculty of Electrical and Computer Engineering, Smart Energy Systems Lab,
University of Tabriz, Tabriz, Iran
e-mail: aliaminlou@tabrizu.ac.ir; mohammadmohsen.hayati@tabrizu.ac.ir



V. Vahidinasab, B. Mohammadi-Ivatloo (eds.), *Demand-Side Peer-to-Peer Energy Trading*, Green Energy and Technology, https://doi.org/10.1007/978-3-031-35233-1_7

electric vehicles (EVs), improve the flexibility of the power system. System end users can also play a significant role in energy system planning by participating in demand response programs (DRPs) [13]. The utilization of DRPs, the emergence of EVs, and the proliferation of integrating DERs in today's power systems change the role of end-users from end-users to both a consumer and a producer simultaneously, called prosumers [14]. As mentioned, the paradigm shift for decarbonization by electric vehicles is realized in both the electricity and transportation sectors, and it is predicted that by 2050, there will be more than one billion electric vehicles (passenger EVs) cars on the world's roads [15]. On the other hand, the significant influence of the presence of electric vehicles for electric sectors with different intelligent charging patterns has created the flexibility potential that electric vehicles can change roles between load and storage devices. It can be said that traditional electricity networks, despite the large number of consumers and producers, are becoming prosumer-centric networks due to the expansion of smart network technologies [16]. Since energy systems are moving towards systems where passive consumers will play a significant role and actively participate in the system with a variety of DERs [17]. In addition to solving the problem of energy balance, the proper design of the energy market will bring widespread social welfare (SW) [4]. For this reason, a mechanism to promote the direct sharing of energy among all multi-level market players with predetermined responsibility and privacy is defined as Peer-to-Peer (P2P) energy trading [18]. The P2P trading mechanism is one of the methods that can be used in local energy markets with goals such as reducing energy costs and peak shaving [19]. This mechanism can facilitate the energy balance from the point of view of the power system locally [20] and is associated with reducing energy transmission losses, improving the reliability of the power system, reducing the supporting electrical infrastructure, and reducing the overall cost of peers [21, 22]. In particular, the Transactive Energy (TE) framework is a new approach to energy management and trading that provides a position for energy trading models to help align between increasing customer choice and participation while respecting the needs of the power system [23]. P2P energy trading under the concept of TE is an optimal and suitable solution for using local DER and loads and coordinating them in modern power systems. In fact, TE enables customers of any size to actively participate in the process of energy trading, consumption, and production [24, 25]. A peer-to-peer energy system is an energy trading mechanism consisting of local energy buyers, sellers, and P2P service providers who come together to trade without the need for intermediaries [5, 26, 27]. This technology can provide system operators with the necessary flexibility to manage grid congestion caused by the lack of coordination of DER operations among buyers and the high intermittency of renewable energy (RE) [26]. P2P energy trading is one of the topics that has become an important research focus in academia and industry. Most of the countries carried out several experimental projects to make this new mechanism effective [17], some of the most important of which are: Piclo in the UK [5, 28], the Brooklyn Microgrid project in the USA [5, 29], and Peer Energy Cloud in Germany [5, 30]. Units participating in the P2P market are called peers. The feasibility of employing P2P trading in a microgrid enables peers to increase the revenue of DER producers and reduce the cost of consumers due to the difference between the selling and buying prices of the network and the P2P market [31].

### 7.1.1 Definition of Transactive Energy Concept

The exact concept of TE first defined by the GridWise Architecture Council is [32] "a system of economic and control mechanisms that allow the dynamic balance of supply and demand across the entire electrical infrastructure using value as a key operational parameter." Transactive energy is a more general form of demand response that can manage production resources in addition to consumption resources. Similar to demand response programs, TE also encourages consumers to exploit their resources according to the needs of the system; With the difference is that transactional energy can extend the concept of demand response DR to production resources and establish the balance between production and consumption in an independent, decentralized, and real-time manner. In other words, TE creates a network of energy sources in which all levels of production and consumption can communicate with each other and exchange energy. TE provides the opportunity to participate in a similar space, called the TE market, for sources of production and consumption that do not have the ability or desire to participate in the wholesale electricity market [33].

### 7.1.2 The General Structure of the P2P Electricity Market

Based on the literature, various structures can be implemented for P2P energy trading. Three types of structures can be proposed for energy trading:

- Centralized P2P trading with an aggregator
- Decentralized P2P trading without an aggregator
- Hybrid P2P trading

### 7.1.3 Centralized Peer-to-Peer Trading

The aggregator, which is a non-profit organization, functions as a local data center for information exchange. The manager of the energy community, who acts as an aggregator, has direct supervision and control over energy import and export restrictions, as well as all energy transactions between sellers and buyers so that buyers can exchange energy through a peer-to-peer trade mechanism. In this model, P2P transactions are cleared with the help of an aggregator [17, 24]. In other words, this aggregator has the ability to provide price signals to peers [9] and manage the energy supply and consumption of peers [34], and after the energy exchange process is completed, it divides the income among buyers according to predetermined criteria [17, 24]. Centralized P2P trading with an aggregator is also called centralized manager-based energy trading [24].

### *7.1.4 Decentralized P2P Trading Without an Aggregator*

In decentralized P2P trading, the participants can directly trade and communicate with each other without the intermediary of the aggregator and the organization called the Energy Community Manager (ECM). Peers in this market have full control over their trading strategies to maximize profits and satisfy personal preferences, as well as all their decision-making processes. Also, the transaction price is at the disposal of the participants. The P2P energy trading market is facilitated by a bilateral contract for transaction information between the energy buyer and seller, which is a simple example of an over-the-counter (OTC) market in energy trading [5, 35].

### *7.1.5 Hybrid P2P Trading*

As all market players can be divided into communities with fewer participants, hybrid markets are more scalable compared to centralized P2P markets. In fact, by reducing the number of players in the market, scalability improves. In these markets, communities trade energy among themselves in a P2P manner, and buyers also trade energy within the community. Of course, there are various challenges in designing a hybrid market structure for energy trading in customer-oriented markets. The first challenge facing hybrid P2P trading is to provide a suitable pricing mechanism to coordinate local trade in the community with neighborhood trade among communities. The second challenge is that, while actors use the network to exchange electricity in different markets, the limitations of the network must be met and considered in energy transactions. Another obstacle in the design process of these markets is the combination of heterogeneous preferences of market players [24, 36].

### *7.1.6 Shared Energy Storage*

Shared energy storage (SES) systems have attracted more attention during the last decade [37]. Due to the rapid development of DERs, (SES) systems have a special role in local communities to ensure the operation of power systems [34, 38]. Shared energy storage systems (SES) are investigated to improve the utilization rate of storage devices and reduce initial investments under the concept of the energy-sharing economy [5]. Energy storage devices such as shared battery storage play an essential role in P2P energy trading markets, enabling peers in this market to store REs and plan their consumption profiles more flexibly. Investigating P2P energy trading markets with the presence of SES has been one of the hot research topics recently [5]. In the P2P trading market, similar to other energy market architectures, different types of DER technologies, such as PV and EV and battery energy storage systems (BESS), can be applied [39].

## 7.2 P2P Energy Market Modeling

Prosumers can participate in two groups of sellers and buyers based on PV energy production and network prices. The peers can sell their extra energy production on the P2P market or sell it to the upper network. This model allows peers to participate in the P2P market individually. Peers are independent players that are trying to maximize their social welfare. The customers can buy a virtual storage capacity from the CSBES to save money by storing PV energy production and discharging it to meet the demand at a higher energy price. The CSBES system shared its daily storage capacity with the peers participating in the P2P market. As shown in Fig. 7.1, several peers are located near each other and connected to a distribution network. Every peer is equipped with a smart meter and secure communication network. Peers can communicate with each other and participate in the P2P market. In this market structure, customers with extra energy production offer their excess energy to the other peers. Customers with energy deficiency meet their demand by buying energy

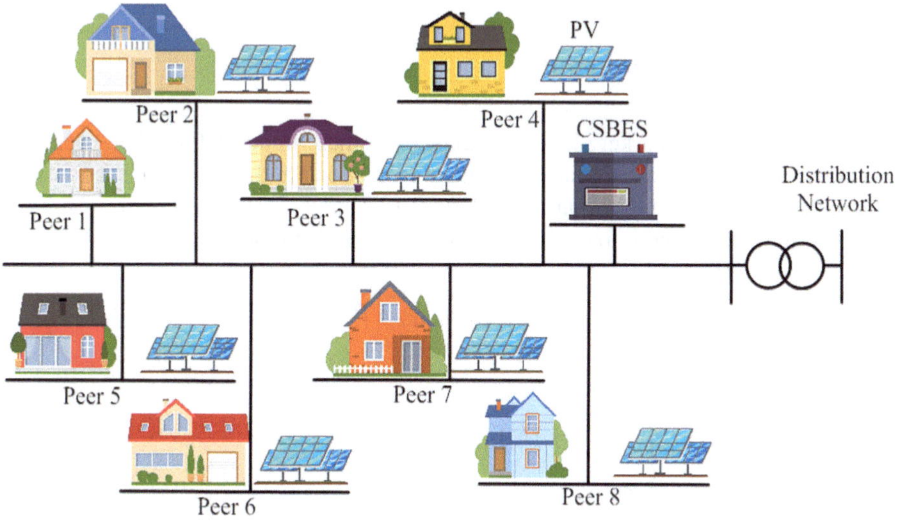


**Fig. 7.1** Schematic overview of a P2P market in an energy community

from the other peers. Trading energy in the P2P market has advantages, such as increasing the reliability of energy supply sources. Also, peers can save their money by joining the P2P market. In the unsubsidized pricing system, the energy selling price of the grid is higher than the buying price. So the peers can earn more money by selling the energy in the P2P market, and the consumers can save more by participating in the P2P market and buying energy from other peers. This modeling solves the P2P market structure with the ADMM algorithm. In this algorithm, the main optimization problem is decomposed into several subproblems, and complicated constraints will be omitted by using the augmented Lagrangian method. ADMM algorithm solves the P2P energy trading problem in several iterations. The ADMM algorithm multipliers update based on the peer's behavior in each iteration. The ADMM multipliers demonstrate the dual variable of the price, editing in a way that reduces the market unbalance. As described before, in the centralized approach for solving the P2P market problem, all information is collected and sent to the market operator. This operator manages the market and solves the optimization problem to minimize the overall cost of the microgrid. This approach does not preserve the peers' information privacy. However, in the decentralized approach where there is no third party to control the market energy management, privacy is maintained. This approach is similar to fundamental markets because participants are intelligent and independent players. This proposed decentralized approach solved the market iteratively until the stopping criteria were approved. In the final iteration, solving results are shared among the peers who track the schedule the following day.

The P2P market structure is formulated in this section with the ADMM algorithm. This model considers the time interval set as $T = \{t_1, t_2, \ldots, t_{24}\}$ that duration of every time slot is one hour. Moreover, the set of $N = \{1, 2, \ldots, 8\}$ indicates the peers participating in the P2P market, and $n$ shows the market participant's number.

### 7.2.1 Objective Function

In this part, the P2P market platform will be examined. In the P2P market, each peer wants to maximize their welfare by managing the load and energy transaction with the upper network and P2P market. Relations (7.1) and (7.2) show the objective and utility function for maximizing the welfare of peers participating in the P2P market.

$$\text{Obj}^{\text{main}} = \max \sum_{t \in T} \sum_{i \in N} \left( P_{\text{S}}^{t} E_{\text{S|G}\ i}^{t} - P_{\text{B}}^{t} E_{\text{B|G}\ i}^{t} - P_{\text{CSBES}}^{t} \text{VSOC}_{i}^{\text{REQ}} + U\left(L_i^t\right) \right) \tag{7.1}$$

$s.\ t$

$$U\left(L_i^t\right) = \begin{cases} \phi_i L_i^t - \gamma_i \left(L_i^t\right)^2 & L_i^t < \dfrac{\phi_i}{2\gamma_i} \\ \dfrac{\phi_i^{\,2}}{2\gamma_i} & L_i^t \geq \dfrac{\phi_i}{2\gamma_i} \end{cases} \tag{7.2}$$

Relation (7.1) shows the main objective function (OF) of the microgrid. This function tries to maximize the social welfare for all of the market. The first part shows the influence of sold energy to the main grid on the OF. The second part demonstrates the bought energy from the main grid. And the third part shows the shared battery energy storage allocation cost. The final part is the utility function described in (7.2). The utility function has three features to show the utility of every customer. This function differs based on the changes in the load. In the first case, zero consumption brings no benefit and the utility function becomes zero by $U(L_i^t) = 0$. In the second case, the first derivative of the utility function is positive by $U'(L_i^t) \geq 0$. In this case, each peer strives to have a stronger grip on energy to assure their welfare. But this increasing welfare does not continue after the saturation level. After the saturation level $U'\ (L_i^t) \leq 0$ reached, consuming more energy brings no advantages to the customers. The behavior of the utility function is dependent on this function's multiplier. This multiplier is defined based on the size and type of the consumers.

$$E_{\mathrm{R}i}^t + \sum_{\substack{j \in N \\ j \neq i}} E_{\mathrm{B|P}\ i,j}^t + E_{\mathrm{B|G}\ i}^t + E_{\mathrm{VDch}\ i}^t = L_i^t + E_{\mathrm{S|G}\ i}^t + \sum_{\substack{j \in N \\ j \neq i}} E_{\mathrm{S|P}\ i,j}^t + E_{\mathrm{VCh}\ i}^t \forall t \in T, i \in N \tag{7.3}$$

$$E_{\mathrm{B|P}\ i,j}^t = E_{\mathrm{S|P}\ j,i}^t \forall t \in T, i \neq j, i, j \in N \tag{7.4}$$

Equation (7.3) demonstrates the energy balance for each peer that participates in the P2P energy trading market. The left side of the equation mentions the incoming energy and the right side shows the outcoming of every peer. The first part of this equation points to the renewable energy production of every peer. This parameter is the forecasted value for the day-ahead scheduling and depends on the sunlight and time of day. $E_{\mathrm{B|G}}^t$ and $E_{\mathrm{B|P}}^t$ demonstrate the energy bought from the upper grid and energy bought from the P2P market. Customers in the P2P market have several choices to meet their demand. One of the traditional ways is buying energy from the upper grid. The second and cleaner way to meet the demand is using the RES and the final way is buying energy in the P2P market. In the balancing equation also, we consider the charging and discharging variable of the virtual energy storage.

Equation (7.4) is the complicated constraint that interconnects the peers to each other. In the centralized approach, this constraint makes sure that consumers bought energy from the producers are the same as the producers sold energy. However, in a decentralized approach by using an ADMM algorithm, this complicated constraint moves to the decomposed objective function of every peer.

$$\mathrm{VSOC}_i^t = \mathrm{VSOC}_i^{t-1} + \eta_{\mathrm{VCh}} E_{\mathrm{VCh}\ i}^t - \left(\frac{1}{\eta_{\mathrm{VDch}}}\right) E_{\mathrm{VDch}\ i}^t \quad \forall t \in T, i \in N \tag{7.5}$$

$$0 \le \mathrm{VSOC}_i^t \le \mathrm{VSOC}_i^{\mathrm{REQ}} \ \forall t \in T, i \in N \tag{7.6}$$

$$\sum_{i \in N} \mathrm{VSOC}_i^{\mathrm{REQ}} \le \mathrm{SOC}^{\mathrm{MAX}} \ \forall t \in T, i \in N \tag{7.7}$$

$$\mathrm{VSOC}_i^{\mathrm{Init}} = 0 \tag{7.8}$$

$$0 \le E_{\mathrm{VCh}\ i}^t \le I_{\mathrm{VCh}\ i}^t E_{\mathrm{VCh}\ i}^{\mathrm{Max}} \ \forall t \in T, i \in N \tag{7.9}$$

$$0 \le E_{\mathrm{VDch}\ i}^t \le I_{\mathrm{VDch}\ i}^t E_{\mathrm{VDch}\ i}^{\mathrm{Max}} \ \forall t \in T, i \in N \tag{7.10}$$

$$I_{\mathrm{VDch}\ i}^t + I_{\mathrm{VCh}\ i}^t \le 1 \quad I_{\mathrm{VDch}\ i}^t, I_{\mathrm{VCh}\ i}^t \in \{0,1\} \ \forall t \in T, i \in N \tag{7.11}$$

As mentioned in the introduction, this model presents a shared battery energy storage system that shares its capacity with the peers that participate in the P2P market. The CSBESS has a limited capacity and this unit rents its capacity to the peers in different time slots in a day. Constraint (7.5) declares that the state of the charge (SOC) for every peer cannot be negative and it should be lower than the requested capacity. Constraint (7.7) will guarantee that the accumulated capacity of individual units does not exceed the boundaries of our main storage unit. In the following, (7.8) shows that the initial SOC value of the virtual capacity is zero. Equations (7.9), (7.10), and (7.11) ensure prevention of the charging and discharging at the same time.

$$E_{\mathrm{B|G}\ i}^t, E_{\mathrm{S|G}\ i}^t, E_{\mathrm{B|P}\ i,j}^t, E_{\mathrm{S|P}\ i,j}^t \ge 0 \ \forall t \in T, i \in N \tag{7.12}$$

$$E_{\mathrm{B|G}\ i}^t + \sum_{\substack{j \in N \\ j \ne i}} E_{\mathrm{B|P}\ i,j}^t \le u_{\mathrm{B}\ i}^t E_{\mathrm{Buy}\ i}^{t\max} \ \forall t \in T, i \in N \tag{7.13}$$

$$E_{\mathrm{S|G}\ i}^t + \sum_{\substack{j \in N \\ j \ne i}} E_{\mathrm{S|P}\ i,j}^t \le u_{\mathrm{S}\ i}^t E_{\mathrm{Sell}\ i}^{t\max} \ \forall t \in T, i \in N \tag{7.14}$$

$$u_{\mathrm{S}\ i}^t + u_{\mathrm{B}\ i}^t \le 1 \quad u_{\mathrm{S}\ i}^t, u_{\mathrm{B}\ i}^t \in \{0,1\} \ \forall t \in T, i \in N \tag{7.15}$$

Equation (7.12) ensure that the P2P and grid energy transaction does not get a negative value. The binary variables $u_{\mathrm{S}}^t, u_{\mathrm{B}}^t$ when taking non-zero value customers can participate in the market as sellers and buyers simultaneously. In following the (7.13), (7.14), and (7.15) do not let the peers arbitrage in the market. For example, this equation prevents the peers from buying energy from the main grid and sell it to the other peers in the P2P market.

### 7.2.2 Employing ADMM on the P2P Market Model

Peers are interconnected with a secure communication line in a microgrid infrastructure. Moreover, there is no third party to manage the energy transaction between the peers in a decentralized approach. This approach increases the complicity in solving the problem. This model employed an ADMM algorithm to decompose the main optimization problem into several sub-problems. In this decomposition, complicated constraints that interconnect subproblems are omitted and sub-problems are solved individually by the peers participating in the P2P market. After all of the peers solve their optimization problem, they update their ADMM multiplier and this process continues until the stopping criteria are provided. In this model for preserving the peer's policy, only the small amount of information signified by a hat is transferred between the peers.

$$\mathrm{Obj}^{i,v} = \max \sum_{t \in T} \left[ \begin{array}{l} P_{\mathrm{S}}^{t} E_{\mathrm{S|G}\ i}^{t} - P_{\mathrm{B}}^{t} E_{\mathrm{B|G}\ i}^{t} - P_{\mathrm{CSBES}}^{t} \mathrm{VSOC}_{i}^{\mathrm{REQ}} + U\left(L_{i}^{t}\right) - \\ \sum_{\substack{j \in N \\ j \neq i}} \lambda_{j,i}^{v,t} E_{\mathrm{B|P}\ i,j}^{v,t} + \lambda_{i,j}^{v,t} \sum_{\substack{j \in N \\ j \neq i}} E_{\mathrm{S|P}\ i,j}^{v,t} \\ -\frac{\rho_{\mathrm{P},i}^{v,t}}{2} \left[ \sum_{\substack{j \in N \\ j \neq i}} \left( E_{\mathrm{B|P}\ i,j}^{t} - \widehat{E}_{\mathrm{S|P}\ j,i}^{t} \right)^{2} + \sum_{\substack{j \in N \\ j \neq i}} \left( E_{\mathrm{S|P}\ i,j}^{t} - \widehat{E}_{\mathrm{B|P}\ j,i}^{t} \right)^{2} \right] \end{array} \right] \quad \forall i \in N \tag{7.16}$$

By applying the augmented Lagrangian to this model, the direct mathematical dependence between the subproblems reform to a penalty value in the objective function. As shown in (7.16), the decomposed subproblems are obtained by using the squared norm of the interconnecting constraint (7.3) and the penalty parameter. The first part of the decomposed OF consists of the utility function and the cost of energy transactions between the peers and the upper grid. The earned money is shown in this function as a positive value. The second part of the decomposed OF shows the penalty parameter. This parameter tends to zero by the convergence process.

$$\mathrm{Obj}^{v} = \sum_{i \in N} \mathrm{Obj}^{i,v} \tag{7.17}$$

The overall OF for the microgrid is obtained by (7.17). In this model, every prosumer solves their optimization problem individually. After each iteration is complete and every prosumer solves their optimization problem, the ADMM multiplier updates using (7.18). This parameter is the dual value of the energy trading price in the P2P market. The initial value of the ADMM multiplier (7.19) is set to the average value of the selling and buying energy from the main grid.

$$\lambda_{i,j}^{v+1,t} = \lambda_{i,j}^{v,t} + \rho_{P,i}^{v,t} \left[ \sum_{\substack{j \in N \\ j \neq i}} E_{\mathrm{B|P}\ j,i}^{v,t} - \sum_{\substack{j \in N \\ j \neq i}} E_{\mathrm{S|P}\ i,j}^{v,t} \right] \forall t \in T, i \in N \tag{7.18}$$

$$\lambda_{i,j}^{0,t} = \left(P_{\mathrm{B}}^{t} + P_{\mathrm{S}}^{t}\right) / 2 \forall i, j \in N, t \in T \tag{7.19}$$

$$r_{\mathrm{P},i}^{v,t} = \sum_{\substack{j \in N \\ j \neq i}} E_{\mathrm{B|P}\ j,i}^{v,t} - \sum_{\substack{j \in N \\ j \neq i}} E_{\mathrm{S|P}\ i,j}^{v,t} \forall t \in T, i \in N \tag{7.20}$$

In the following, the residual value shown in (7.20) provides a perspective from the convergence of the Algorithm in the P2P market. This value is equal to the difference between the overall sold energy and bought energy in the P2P market. If the residual value approach to the $\varepsilon$ which is considered as 10W the iterations stop and the final results of day-ahead scheduling are shared between the prosumers.

## 7.3 Simulation Results

As shown in Fig. 7.1, eight peers are connected to a distribution network under a secure communication link. They participate in the energy market based on the network price and PV energy production with the presence of CSBES. The results of this study with a decentralized approach show that energy transaction management between peers reduces the overall cost.

### *7.3.1 Implementation of the Proposed Model and Data Analysis*

In this research, a microgrid consisting of peers equipped with smart meters to measure energy transactions and a shared central battery storage unit is considered to investigate the effectiveness of the proposed decentralized approach. In this study, the proposed ADMM model has been implemented on GAMS software (V24.9.1) and solved with the Mixed integer nonlinear programming (MINLP) with BARON solver. Also, all peers in this microgrid have distributed energy resources except the first unit. As shown in Fig. 7.2, the PV unit has different installed power, which is from 2 to 8 peers, respectively. In addition, according to Fig. 7.3, the results show

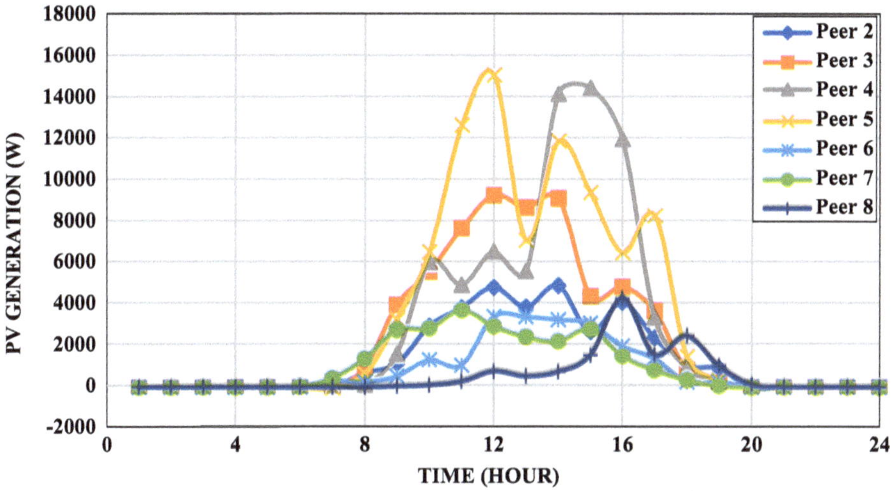


**Fig. 7.2** PV Generation Profile of each peer

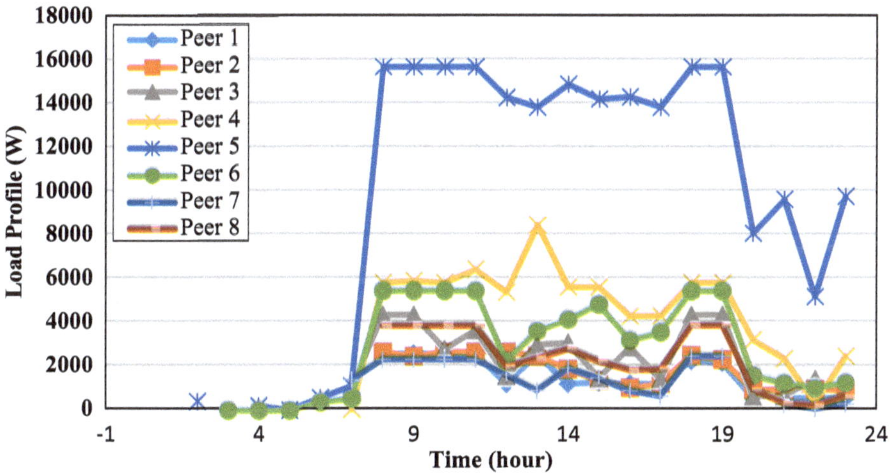


**Fig. 7.3** Load profile of each peer

that the load profile of each peer has different types, such as science buildings and residential buildings, for the results to be closer to reality.

PV generation and load characteristics are among the factors that directly affect customer behavior in the P2P market, and excess PV energy is sold in the P2P market. In this study, except for the first peer, other peers are equipped with a photovoltaics array, and all peers are directly connected to the distribution network and PV generation and load specifications were adopted from [40].

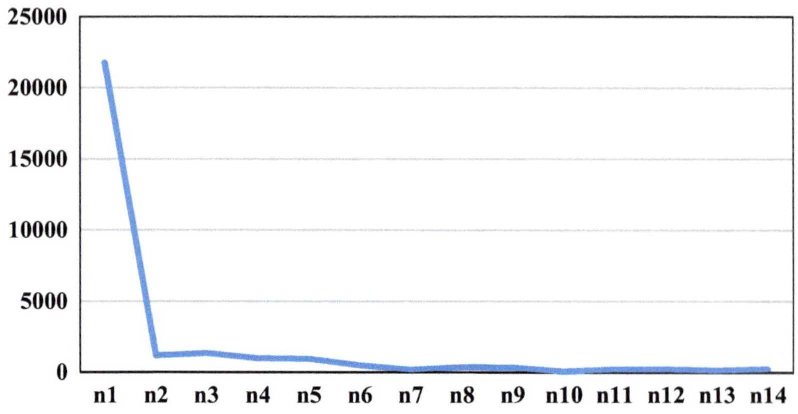


**Fig. 7.4** The number of iterations convergence profile of ADMM algorithm

In this paper, peers cooperate and the impact of decentralized P2P energy trading is evaluated on several customers' profiles. The discussed simulations are run over 24 hours, divided into one-hour time intervals. In addition, the results of the simulations indicate that the peers try to transfer the load from the peak time to the non-peak time. The price of P2P energy trading in the first iteration is set to the average of the network sale and purchase price. Also, in the next iteration, to reduce the energy imbalance, the P2P market price is updated by using the ADMM algorithm. The main grid used the Time of Use (TOU) pricing scheme, and the P2P price will be lower than the network price. We know that in the TOU pricing plan, the network price is divided into several fixed time slots, and in each one, the network price differs based on the time.

As shown in Fig. 7.4, the first iterations show the convergence rate, but the P2P market does not achieve the convergence condition until the 14th iteration. Regardless of the delay in solving the optimization problem and other limitations in the communication network platform, it takes about five seconds for each iteration. Figure 7.5 shows the total state of charge and also shows the SOC profile of each peer in the presence of CSBES in one day. The P2P trading market provides extra income for sellers and this market is more economical for buyers. Analyzing the results, the results show that decentralized P2P energy trading with the presence of CBESS reduces the overall cost.

In the following, the Figs. 7.6 and 7.7 show the overall energy that bought or sold from the main grid.

Figures 7.8 and 7.9 contain an important information about the peer-to-peer energy trading. This result shows the energy transaction between the peers in several times of a day. In these figures, we can observe that the sold energy of peer $i$ to the peer $j$ is equal to the energy that peer $j$ bought from unit $i$.

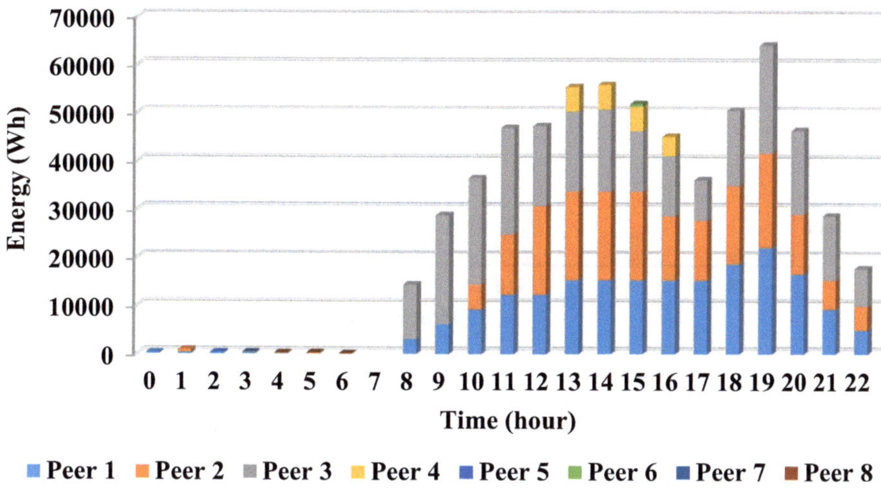


**Fig. 7.5** State of charge (SOC) of each peer in presence of CSBES

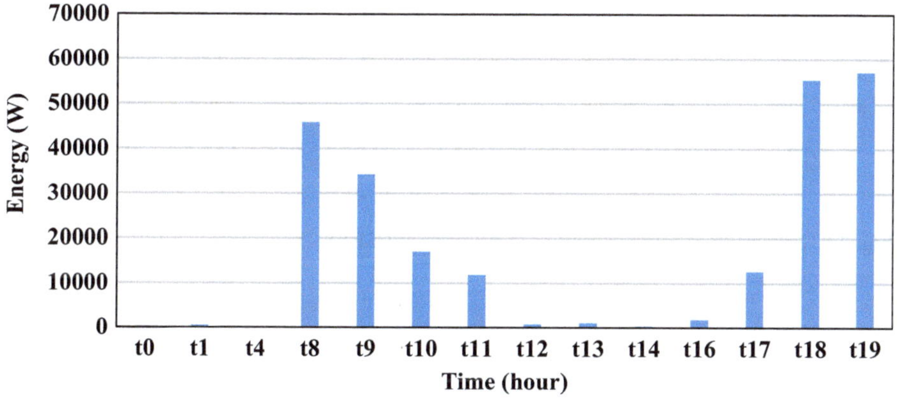


**Fig. 7.6** Schematic of overall buy from the grid

## 7.4 Conclusions

This chapter discussed the P2P energy trading concept as a promising solution for decreasing the cost of the customers. Also, the related literature, in this regard, was reviewed. This concept also increases energy efficiency by reducing the distance between producers and consumers and increases reliability by increasing the energy supply sources. The P2P energy trading concept intensifies the prosumers to take part in a P2P market. This new market structure paves the way for other emerging technologies, like central battery energy storage. The CBESS is an integrated facility

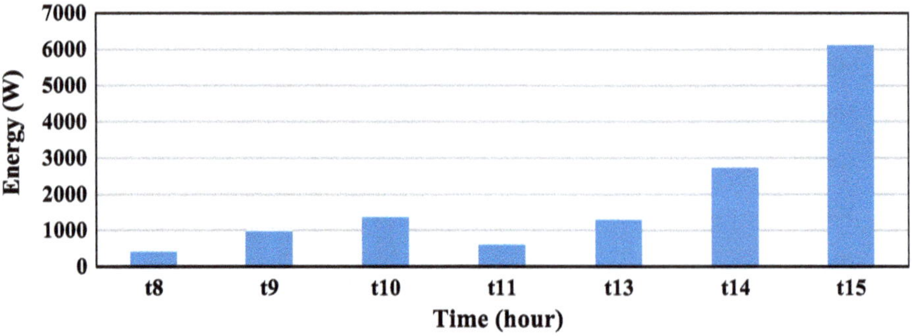


**Fig. 7.7** Schematic of overall sell to the grid

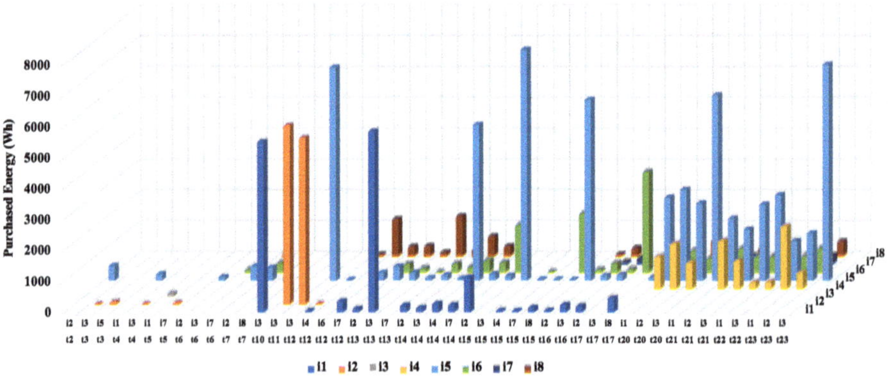


**Fig. 7.8** Schematic of energy purchase in presence of CSBES

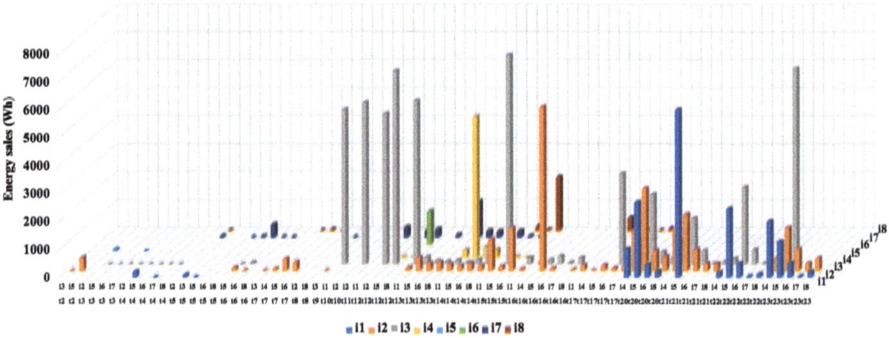


**Fig. 7.9** Schematic of Energy sales in the presence of CSBES

that contains several battery units. This unit has a huge capacity for storing energy at a particular time and discharging it when the system needs that energy injection. The CBESS shares its capacity with the customers who participate in the P2P market in exchange for a fee. In this market structure, peers are interconnected with a secure communication link, and they can contact each other and trade energy. This chapter solves the market clearing problem with a decentralized approach that cause to preserve the peer's information privacy. The ADMM algorithm is employed for modeling this market. Simulation results show that every peer tries to maximize their social welfare by participating in the P2P market and they use the SBES to reduce their costs by charging in the off-peak time and discharging in the peak times.

## References


1. Kalantari, H., Sasmito, A. P., & Ghoreishi-Madiseh, S. A. (2021). An overview of directions for decarbonization of energy systems in cold climate remote mines. *Renewable and Sustainable Energy Reviews, 152*, 111711.
2. Chenic, A. Ş., et al. (2022). Logical analysis on the strategy for a sustainable transition of the world to green energy—2050. Smart cities and villages coupled to renewable energy sources with low carbon footprint. *Sustainability, 14*(14), 8622.
3. Aguero, J. R., Takayesu, E., Novosel, D., & Masiello, R. (2017). Modernizing the grid: Challenges and opportunities for a sustainable future. *IEEE Power and Energy Magazine, 15*(3), 74–83.
4. Liu, C., & Li, Z. (2022). Comparison of centralized and peer-to-peer decentralized market designs for community markets. *IEEE Transactions on Industry Applications, 58*(1), 67–77.
5. Zheng, B., Wei, W., Chen, Y., Wu, Q., & Mei, S. (2022). A peer-to-peer energy trading market embedded with residential shared energy storage units. *Applied Energy, 308*, 118400.
6. Aminlou, A., Mohammadi-Ivatloo, B., Zare, K., Razzaghi, R., & Anvari-Moghaddam, A. (2022). Peer-to-peer decentralized energy trading in industrial town considering central shared energy storage using alternating direction method of multipliers algorithm. *IET Renewable Power Generation, 16*, 2579–2589.
7. Rissman, J., et al. (2020). Technologies and policies to decarbonize global industry: Review and assessment of mitigation drivers through 2070. *Applied Energy, 266*, 114848.
8. Tsao, Y.-C., Vu, T.-L., & Lu, J.-C. (2021). Pricing, capacity and financing policies for investment of renewable energy generations. *Applied Energy, 303*, 117664.
9. Chen, Y., Wei, W., Wang, H., Zhou, Q., & Catalão, J. P. (2021). An energy sharing mechanism achieving the same flexibility as centralized dispatch. *IEEE Transactions on Smart Grid, 12*(4), 3379–3389.
10. Helistö, N., et al. (2019). Backbone—An adaptable energy systems modelling framework. *Energies, 12*(17), 3388.
11. Chen, Y., Lei, X., Yang, J., Zhong, H., & Huang, T. (2022). Decentralized P2P power trading mechanism for dynamic multi-energy microgrid groups based on priority matching. *Energy Reports, 8*, 388–397.
12. Mehdinejad, M., Shayanfar, H., & Mohammadi-Ivatloo, B. (2022). Peer-to-peer decentralized energy trading framework for retailers and prosumers. *Applied Energy, 308*, 118310.
13. Kahnamouei, A. S., & Lotfifard, S. (2022). Enhancing resilience of distribution networks by coordinating microgrids and demand response programs in service restoration. *IEEE Systems Journal, 16*(2), 3048–3059.
14. Da Silva, P. G., Ilić, D., & Karnouskos, S. (2013). The impact of smart grid prosumer grouping on forecasting accuracy and its benefits for local electricity market trading. *IEEE Transactions on Smart Grid, 5*(1), 402–410.

15. Borray, A. F. C., Merino, J., Torres, E., & Mazón, J. (2020). A review of the population-based and individual-based approaches for electric vehicles in network energy studies. *Electric Power Systems Research, 189*, 106785.
16. Mehdinejad, M., Shayanfar, H., & Mohammadi-Ivatloo, B. (2022). Decentralized blockchain-based peer-to-peer energy-backed token trading for active prosumers. *Energy, 244*, 122713.
17. Tushar, W., et al. (2021). Peer-to-peer energy systems for connected communities: A review of recent advances and emerging challenges. *Applied Energy, 282*, 116131.
18. Wu, Y., Wu, Y., Cimen, H., Vasquez, J. C., & Guerrero, J. M. (2022). P2P energy trading: Blockchain-enabled P2P energy society with multi-scale flexibility services. *Energy Reports, 8*, 3614–3628.
19. Huang, Y., Valipour, E., Nojavan, S., Hoseinzadeh, M., & Kargar, S. (2022). Risk modeling of an industrial area with shared storage of several buildings using downside risk constraint method considering peer-to-peer trading. *Journal of Energy Storage, 47*, 103540.
20. Saatloo, A. M., Mirzaei, M. A., & Mohammadi-Ivatloo, B. (2022). A robust decentralized peer-to-peer energy trading in community of flexible microgrids. *IEEE Systems Journal, 99*, 1–12.
21. Zia, M. F., Benbouzid, M., Elbouchikhi, E., Muyeen, S., Techato, K., & Guerrero, J. M. (2020). Microgrid transactive energy: Review, architectures, distributed ledger technologies, and market analysis. *IEEE Access, 8*, 19410–19432.
22. Van Cutsem, O., Dac, D. H., Boudou, P., & Kayal, M. (2020). Cooperative energy management of a community of smart-buildings: A Blockchain approach. *International Journal of Electrical Power & Energy Systems, 117*, 105643.
23. Daneshvar, M., Mohammadi-Ivatloo, B., Zare, K., & Asadi, S. (2021). Transactive energy management for optimal scheduling of interconnected microgrids with hydrogen energy storage. *International Journal of Hydrogen Energy, 46*(30), 16267–16278.
24. Khorasany, M., Mishra, Y., & Ledwich, G. (2020). Hybrid trading scheme for peer-to-peer energy trading in transactive energy markets. *IET Generation, Transmission & Distribution, 14*(2), 245–253.
25. Morstyn, T., Farrell, N., Darby, S. J., & McCulloch, M. D. (2018). Using peer-to-peer energy-trading platforms to incentivize prosumers to form federated power plants. *Nature Energy, 3*(2), 94–101.
26. Esmat, A., de Vos, M., Ghiassi-Farrokhfal, Y., Palensky, P., & Epema, D. (2021). A novel decentralized platform for peer-to-peer energy trading market with blockchain technology. *Applied Energy, 282*, 116123.
27. Mehdinejad, M., Shayanfar, H. A., Mohammadi-Ivatloo, B., & Nafisi, H. (2022). Designing a robust decentralized energy transactions framework for active prosumers in peer-to-peer local electricity markets. *IEEE Access, 10*, 26743–26755.
28. Piclo. (2021). Available from: https://piclo.energy/. Accessed 28 Jan 2021.
29. Mengelkamp, E., Gärttner, J., Rock, K., Kessler, S., Orsini, L., & Weinhardt, C. (2018). Designing microgrid energy markets: A case study: The Brooklyn microgrid. *Applied Energy, 210*, 870–880.
30. Peer Energy Cloud. (2021). Available from: https://software-cluster.org/projects/peer-energy-cloud/. Accessed 19 Jan 2021.
31. Long, C., Wu, J., Zhang, C., Thomas, L., Cheng, M., & Jenkins, N. (2017). Peer-to-peer energy trading in a community microgrid. In *2017 IEEE power & energy society general meeting* (pp. 1–5). IEEE.
32. T. Council. (2015). Gridwise transactive energy framework version 1.0, *The GridWise Architecture Council, Tech. Rep.*
33. Rayati, M., Amirzadeh Goghari, S., Nasiri Gheidari, Z., & Ranjbar, A. M. (2018). Robust and dynamic transactive energy system using Tsypkin–Polyak theorem. *IET Smart Grid, 1*(2), 57–65.
34. Long, C., Wu, J., Zhou, Y., & Jenkins, N. (2018). Peer-to-peer energy sharing through a two-stage aggregated battery control in a community microgrid. *Applied Energy, 226*, 261–276.

35. Sorin, E., Bobo, L., & Pinson, P. (2018). Consensus-based approach to peer-to-peer electricity markets with product differentiation. *IEEE Transactions on Power Systems, 34*(2), 994–1004.
36. Sousa, T., Soares, T., Pinson, P., Moret, F., Baroche, T., & Sorin, E. (2019). Peer-to-peer and community-based markets: A comprehensive review. *Renewable and Sustainable Energy Reviews, 104*, 367–378.
37. Dai, R., Esmaeilbeigi, R., & Charkhgard, H. (2021). The utilization of shared energy storage in energy systems: A comprehensive review. *IEEE Transactions on Smart Grid, 12*(4), 3163–3174.
38. Lüth, A., Zepter, J. M., del Granado, P. C., & Egging, R. (2018). Local electricity market designs for peer-to-peer trading: The role of battery flexibility. *Applied Energy, 229*, 1233–1243.
39. Zhang, Z., Tang, H., Wang, P., Huang, Q., & Lee, W.-J. (2019). Two-stage bidding strategy for peer-to-peer energy trading of nanogrid. *IEEE Transactions on Industry Applications, 56*(2), 1000–1009.
40. PVOutput. https://pvoutput.org/, Accessed 20 Dec 2022.

# Chapter 8
# Robust Optimization-Based Decentralized Peer-to-Peer Energy Trading for Prosumers

**Mohammad Seyfi, Mostafa Yaghoubi, Mehdi Mehdinejad, Zbigniew Dziong, and Heidarali Shayanfar**

## 8.1 Introduction

The expansion of smart grid infrastructure and communication technologies, along with the widespread deployment of distributed energy resources (DERs), has made electricity markets face many challenges. Because passive consumers have become consumers with the capability to generate and load management and have formed a new concept called the prosumer [1]. With the advent of prosumers and their desire to trade energy with each other, local electricity markets were formed, and a concept called peer-to-peer energy trading (P2P) was created. With the formation of these new markets, electricity markets changed from their conventional hierarchical structure to a decentralized prosumer-centric model [2]. In the prosumer-centric model, each prosumer trades their surplus energy with other prosumers who have deficient energy [3]. In this regard, by defining appropriate operation mechanisms, it can trade energy with the upstream (retail) market and obtain maximum economic benefits [4]. According to the capabilities provided by the smart grid infrastructure, prosumers can participate in various demand response (DR) programs without relying on a central institution and according to their own preferences. To manage the uncertainty of the upstream market price [5] and its risks, they can use electric storage technology to properly manage their load and generation and optimize their objective function.

M. Seyfi · M. Mehdinejad (✉) · H. Shayanfar
Center of Excellence for Power Systems Automation and Operation, School of Electrical Engineering, Iran University of Science and Technology, Tehran, Iran

M. Yaghoubi · Z. Dziong
Department of Electrical Engineering, Ecole de Technologie Superieure (ETS), Montreal, QC, Canada
e-mail: mostafa.yaghoubi.1@ens.etsmtl.ca; zbigniew.dziong@etsmtl.ca

 
V. Vahidinasab, B. Mohammadi-Ivatloo (eds.), *Demand-Side Peer-to-Peer Energy Trading*, Green Energy and Technology, https://doi.org/10.1007/978-3-031-35233-1_8

The clearing approach is one of the major challenges for peer-to-peer markets. Due to the decentralized nature of these markets, the clearing approach should maintain the decentralized feature to allow each player to individually optimize their problem. The clearing approaches of electricity markets can be different depending on the type of structure, market assumptions, particular rules, and the players' behaviors. In general, peer-to-peer market clearing approaches are divided into two categories: centralized and distributed. In the centralized approach, all decision-making data of the players are aggregation and integrated by a central operator [6, 7]. In this type of market clearing approach, preserving the privacy of the decision-making information of players is very difficult or not possible. But in distributed approaches, each player is trying to solve its own optimization problem with minimal information from other players. Various distributed methods such as the primal-dual gradient method [8–10], alternating direction method of multipliers [11–15], the fast alternating direction method of multipliers [16, 17], consensus-based methods [18, 19], and decentralized ant-colony optimization [20] have been used to clear these markets to date.

### *8.1.1 Related Works*

The studies undertaken related to this chapter's main idea have been evaluated in this section. Consequently, the studies were evaluated based on a number of factors, including market clearing, participation in the upstream market, proposed market models (centralized or decentralized), robust optimization, penetration of electric storage among prosumers, participation in demand response programs, and managing the uncertainty of the upstream market price.

The authors of [10] developed a P2P energy trading model for a small distribution network without an upstream market connection. To clear the proposed decentralized market model, they have used the primal-dual gradient approach. The authors in [21] have proposed a peer-to-peer market model for the transmission level. However, their proposed model is not fully decentralized. Fully decentralized models are those in which each of the buyers and sellers conducts bilateral negotiations with each other. However, in the proposed model, sellers cannot select their energy peers and only act as price-makers. In order to clear the market, the primal-dual gradient approach has been used, which is not a comprehensive approach because it is only applicable to convex quadratic models [22]. In [23], the authors developed a network of bilateral contracts that encourages peer-to-peer energy trading among all players in the power system, such as large producers, intermediary providers, renewable sources, and consumers with flexible loads. In the proposed model, the uncertainty of the market price and participation in demand response programs haven't been considered. The authors in [24] designed a market based on peer-to-peer trading in the presence of retailers and prosumers. In the proposed model, retailers play the role of an energy peer for prosumers. To clear the proposed decentralized market, the authors have used the primal-dual sub-gradient approach.

In [25], the authors have proposed a P2P market for prosumers in a community in which home battery storage systems and shiftable home appliances are used to facilitate energy trading. The authors believe that their proposed model is decentralized and hierarchical. However, the trading of energy with the upstream market and the effect of its price uncertainty in the proposed market is not considered.

A peer-to-peer energy-backed token market is proposed in [26]. The proposed model is fully decentralized, and prosumers can participate in the demand response programs. Prosumers can also trade energy with the upstream market without considering its price uncertainty (the upstream market price). This proposed market has been cleared using a primal-dual sub-gradient approach.

The authors in [27] have designed an energy market based on peer-to-peer trading between virtual power plants with the help of smart contracts under blockchain technology. Upstream market price uncertainty is ignored in the proposed model. In [28], the authors proposed a stable-matching (SM) algorithm to help prosumers find their closest energy peers. This algorithm acts as an operator of the distribution system and identifies the shortest electric route between prosumers. This algorithm acts as a distributed system operator and identifies the shortest electric route between prosumers. The proposed energy market is cleared by using a continuous double auction (CDA) after it has been determined which energy peers are closest to each other. In [29], the authors designed a peer-to-peer energy market within a community of prosumers. This community has been proposed as a virtual power plant. The smart contract on the Ethereum blockchain platform has been used in this design. The proposed market has been cleared using the auction mechanism in the blockchain platform. In [20], a peer-to-peer energy market with parallel auction and pool structured for prosumers using blockchain technology has been designed and cleared using decentralized ant-colony optimization. In [30], the authors have proposed an optimal routing to avoid line congestion in peer-to-peer energy trading in a distribution network. In addition, the proposed market was cleared using a metaheuristic optimization algorithm.

Given the performed studies, the following study gaps are evident:

1. *Fully decentralized peer-to-peer energy market model*: Many of the P2P models proposed are not fully decentralized. Thus, either all players do not negotiate bilaterally, or they rely on a central institution. For example, [7, 21, 25, 28, 30].
2. *Energy transaction of prosumers with the upstream network and considering uncertainty in upstream market price*: In most of the proposed models, prosumers do not participate in the upstream market and only participate in the local market. In the studies where the prosumers participate in the upstream market, the prosumers are prevented from free participation in this market by the prescriptive pricing. For example, [10, 18, 20, 21, 25, 28, 30]. In practice, the uncertainty of the upstream market price has a great influence on the planning of prosumers. By considering a more realistic market behavior, the proposed model can show more robustness in the face of uncertainties as a stochastic event.

3. *Free participation in the demand response program*: In order to participate in demand response programs, prosumers must make independent decisions. In some studies, prosumers do not participate in demand response programs (not modeled). In others, an operator decides on the participation of prosumers in demand response programs. For example, [10, 18, 20, 21, 25, 28, 30].
4. *Market clearing approaches*: In order to clear a fully decentralized market, a completely decentralized approach must be used. In many studies, centralized approaches have been used to clear markets that have a decentralized nature.

Considering the mentioned study gaps, this chapter focuses on developing a novel fully decentralized P2P energy trading market for small-scale prosumers. In this market, all market players are free to determine their desired trade and energy price. Furthermore, this market omits the role of supervisory nodes which results in the privacy protection of market players. Moreover, in this chapter, the decentralized demand response program is integrated with the local P2P energy trading market. Buyers of this market are able to shift their flexible load to manage their hourly demand and reduce their costs.

In addition, the uncertainty of the energy price is considered in this chapter. The robust optimization is used to model the uncertainties of the energy price and investigate the alterations in the player's strategies caused by changes in the degree of conservatism.

A new fast ADMM method is developed for this market model to clear the trades between prosumers. This method has a better convergence rate than the former versions of the ADMM approach.

## 8.2 Problem Formulation

### *8.2.1 P2P Energy Trading Market Characteristics*

Designing a competitive and free market for prosumers needs a proper method and mathematical modeling. Prosumers are greedy agents searching for the best opportunities in the market to increase their welfare, which mainly includes the economic aspects. However, other targets, like expanding the use of clean energy, are possible for prosumers. Therefore, it seems necessary to create a market structure which can provide free platform for prosumers to use and decide on their trades. More importantly, the privacy protection of prosumers emphasizes the need for a free and decentralized market, which makes them able to negotiate and trade by their own decision. Generally, in competitive markets, the specifications of market players should be hidden from others to prevent abusing this information.

Considering these points, this chapter provides a fully decentralized P2P energy trading market for active prosumers. In this market, all market players are responsible for their activities and energy transactions. In other words, prosumers determine their traded energy with their peers and its price. They will not make a deal with other prosumers until their objectives are guaranteed. Overall, they do not care about the social welfare and other players' objective function.

The developed P2P energy trading market includes local buyers and sellers. While prosumers can buy and sell energy simultaneously, for simplicity, we divided them into two groups of buyers and sellers. The proposed model can be upgraded to consider the dual role of prosumers in the market. Buyers are prosumers who need electrical energy to fulfill their energy demands. On the other hand, prosumers who have electrical energy for selling in the P2P energy trading market are called sellers. These roles can be altered at different hours and on different days. In other words, the generation net of prosumers determines their role in the decentralized markets. We consider that sellers are equipped with PVs, which makes them able to produce energy when solar energy is enough.

Buyers as end consumers can participate in the demand response program to enable demand-side management. It is considered that in this market, buyers have a flexible demand and can shift some portion of their hourly demand during the 24 hours of the day. This is known as the load shifting-demand response program. This mechanism helps buyers manage their demand and reduce the cost of purchasing energy. They can shift their demands to cheaper hours in order to consume energy at lower prices. Furthermore, buyers in this market can shift the bought energy using their energy storage system. Using this storage, buyers can buy their demand at low prices and charge their storage, then discharge this energy when their demand is high or when the energy price in the energy market isn't affordable.

In this market, buyers and sellers negotiate among themselves to discover the peer that meets their demands and objectives. They can have multiple peers at the same time and divide the buying or selling energy to various trades. Therefore, the competition in the market determines the percentages of total energy that prosumers trade with each peer. In these markets, the energy price plays the most important role. Because financial welfare is the main goal of prosumers, buyer desire to find the cheapest deals, and on the other hand, sellers attempt to sell their produced energy at the maximum price.

The developed local P2P is connected to the upstream day-ahead markets. Therefore, the market in this market can heavily affect the activities of market players in the local market. Prosumers in local market are free agents and if they find the upstream market more affordable, they can participate in this market. As a result, the hourly price in local market will follow the price profile in the upstream market. It means that the average hourly price in the local market is close to the hourly energy price in the upstream market.

Accordingly, it is necessary to determine the energy price in the upstream market. However, it is hard to completely predict this value. It means that there is a big uncertainty in the value of the hourly energy price. In this chapter, the robust model is utilized to model the uncertain behavior of energy price. This method follows a conservative procedure in which the worst-case scenario is considered. Thus, all other solutions that could be obtained will be better than the solutions found by this method.

As aforementioned, prosumers are divided into two groups: sellers and buyers. If $N_{\mathrm{B}} = \{1, 2, .., n_{\mathrm{b}}\}$ is considered as the buyers set and $N_{\mathrm{s}} = \{1, 2, .., n_{\mathrm{s}}\}$ as the sellers set, then $N_{\mathrm{B}} \cap N_{\mathrm{s}} = \varnothing$. It means that at each time slot, a prosumer can only be either a buyer or a seller.

### 8.2.2 Sellers Model

The main objective of sellers in the developed local P2P energy trading market is to gain the maximum profit by selling their produced energy at the highest price. Therefore, the objective function of sellers can be written as (8.1), where there are two terms: the profit of sold energy and the cost of producing energy. The total energy produced by sellers can be calculated by (8.2), which is a summation of energy sold to all other players. Equation (8.3) states the maximum and minimum limitations of energy that can be produced by sellers. The cost function of sellers is a quadratic function, and its value depends on the total energy produced by sellers. Equation (8.4) states the cost function of sellers. Also, Eq. (8.5) states that the sold energy to all other players should be greater than zero.

$$\max_{x_i} \mathrm{WS}_i = \sum_{t=1}^{T} \left( \sum_{j=1}^{N_{\mathrm{B}}} \lambda_{ijt} x_{ijt} - C(x_{it}) \right) \tag{8.1}$$

s.t. :

$$x_{it} = \sum_{j=1}^{N_{\mathrm{B}}} x_{ijt} \tag{8.2}$$

$$x_{it}^{\min} \leq x_{it} \leq x_{it}^{\max} \tag{8.3}$$

$$C(x_{it}) = \alpha_i x_{it}^2 + \beta_i x_{it} + \gamma_i \tag{8.4}$$

$$x_{ijt} \geq 0 \tag{8.5}$$

### 8.2.3 Buyers Model

The first goal of buyers for participating in the local P2P energy trading market is to buy the required energy to provide their demand. Therefore, a utility function should be defined in order to model their behavior and welfare when consuming electrical energy. This function should have an increasing value per kWh increase on the consumed energy. One utility function for this purpose has been introduced in [31], which is used in this chapter. This is a quadratic function that can model the desire of buyers to consume electrical energy.

$$U(y_{jt}) = \begin{cases} \omega_j y_{jt} - \delta_j {y_{jt}}^2 & y_{jt} < \dfrac{\omega_j}{2\delta_j} \\ \dfrac{{\omega_j}^2}{2\delta_j} & y_{jt} \geq \dfrac{\omega_j}{2\delta_j} \end{cases} \tag{8.6}$$

By using this function, the welfare of buyers can be written as:

$$\max_{y_j} \mathrm{WB}_j = \sum_{t=1}^{T} \left( U(y_{jt}) - \sum_{i=1}^{\mathcal{N}_S} \lambda_{ijt} y_{jit} - \lambda_t^R y_{jt}^g \right) \tag{8.7}$$

In order to complete the optimization model of buyers, their constraints should be added to the objective function. Equation (8.8) states the power balance constraint of buyers, where the hourly consumption of buyers should be equal to the summation of energy bought from local sellers, energy bought from the upstream network, the energy shifted through the demand response program, and charge and discharge of the energy storage system. The maximum amount of energy that can be shifted by buyers in the demand response program is stated by (8.9). According to (8.10), the total daily energy shifted by buyers should be zero. In other words, the demand response program cannot remove some part of the demand of buyers, but it can shift it from 1 hour to another. As mentioned before, buyers have their own energy storage system and can use it to reduce their energy consumption costs. The energy stored in the energy storage system at the end of each hour can be calculated by (8.11). This value depends on the remaining energy at the end of the previous hour, the charged energy at the current hour, and the discharged energy at the current hour. Equations (8.12)–(8.15) state the other constraints of energy storage systems, including the maximum and minimum limitations of stored energy, maximum charge and discharge rate of storage, and the binary variables for defining the operational state of energy storage systems.

$$y_{jt} = \sum_{i=1}^{\mathcal{N}_S} y_{jit} + y_{jt}^g + y_{jt}^{\mathrm{dch}} - y_{jt}^{\mathrm{ch}} - y_{jt}^{\mathrm{DR}} \tag{8.8}$$

$$-DR^{\max} D_{jt} \leq y_{jt}^{DR} \leq DR^{\max} D_{jt} \tag{8.9}$$

$$\sum_{t=1}^{T} y_{jt}^{\mathrm{DR}} = 0 \tag{8.10}$$

$$E_{jt} = E_{jt-1} + \eta^{\mathrm{ch}} \eta^{\mathrm{ch}} y_{jt}^{\mathrm{ch}} - \frac{y_{jt}^{\mathrm{dch}}}{\eta^{\mathrm{dch}}} \tag{8.11}$$

$$0 \leq y_{jt}^{\mathrm{ch}} \leq b_{jt}^{\mathrm{ch}} y_j^{\mathrm{ch,max}} \tag{8.12}$$

$$0 \leq y_{jt}^{\text{dch}} \leq b_{jt}^{\text{dch}} y_j^{\text{dch,max}} \tag{8.13}$$

$$E_j^{\min} \leq E_{jt} \leq E_j^{\max} \tag{8.14}$$

$$b_{jt}^{\text{ch}} + b_{jt}^{\text{dch}} \leq 1 \tag{8.15}$$

$$y_{jit}, y_{jt}^{\text{ch}}, y_{jt}^{\text{dch}}, E_{jt} \geq 0, b_{jt}^{\text{ch}}, b_{jt}^{\text{dch}} \in \{0, 1\} \text{ and } y_{jt}^{\text{DR}} \text{ is free} \tag{8.16}$$

### 8.2.4 Optimization Problem of P2P Market

The total objective function of the developed P2P energy trading market is the summation of the objective functions of buyers and sellers. This is the objective function of the central model of the energy trading market, where social welfare is considered in the mathematical modeling of the optimization problem. The constraints of this problem are all constraints related to buyers and sellers.

$$\max_{x,y} \left( \sum_{i=1}^{\mathcal{N}_S} \text{WS}_i + \sum_{j=1}^{\mathcal{N}_B} \text{WB}_j \right) \tag{8.17}$$

When it comes to the main objective function of this market, there is another constraint which is called the coupling constraint. In other words, this constraint relates the optimization problems of sellers and buyers together. According to (8.18), at each hour, the energy bought by buyer $j$ from seller $i$ should be equal to the energy sold by the seller $i$ to buyer $j$.

$$x_{ijt} = y_{jit} : \lambda_{ijt} \tag{8.18}$$

### 8.2.5 Uncertainty Modeling

Generally, uncertainties are an important part of optimization problems. They exist because it is almost impossible to predict the parameters with no errors. In the developed market in this chapter, the energy price in the upstream network has an uncertain nature, and it can change during a small period. It is hard for buyers to forecast it for buyers in order to decide on their transactions in the market. Therefore, the uncertainty of the energy price should be considered in the optimization problems related to energy markets.

In this chapter, the robust approach is used to model the uncertainty of the energy price. This method is used when limited information about statistical specifications of the uncertain parameter exist. In the robust approach, the uncertainty set is used to define the alteration interval for the uncertain parameter. We use the $U$ uncertainty set for modeling the uncertainty of electrical energy price.

The $U$ uncertainty set can be defined as follows:

$$\mathcal{U}_\lambda = \left\{ \tilde{\lambda}_t^R \in R^+ : \underline{\Gamma}_\lambda \le \frac{\sum \tilde{\lambda}_t^R}{\sum \bar{\lambda}_t^R} \le \bar{\Gamma}_\lambda \forall t, \tilde{\lambda}_t^R \in \left[\lambda_t^{R,\min}, \lambda_t^{R,\max}\right] \right\} \tag{8.19}$$

In this equation, $\lambda_t^{R,\min}$ and $\lambda_t^{R,\max}$ are the maximum and minimum values for the variation of uncertain parameter $\tilde{\lambda}_t^R$. These limits are applied to the uncertain parameter using $\Gamma$, which is the budgeting coefficient. $\bar{\Gamma}_\lambda$ and $\underline{\Gamma}_\lambda$ define how conservative is the problem solution.

The uncertain parameter exists in the objective function of the optimization problem. It can be rewritten as follows:

$$\begin{gathered} \max_{y_j} \mathrm{WB}_j = z_j \\ \text{S.t.} \end{gathered} \tag{8.20}$$

$$z_j + \sum_{t=1}^{T} \tilde{\lambda}_t^R y_{jt}^g \le \sum_{t=1}^{T} U(y_{jt}) - \sum_{t=1}^{T} \sum_{i=1}^{\mathcal{N}_S} \lambda_{it} y_{jit} \tag{8.21}$$

The uncertain parameter and its variation from the nominal value can be stated as follows:

$$\tilde{\lambda}_t^R = \bar{\lambda}_t^R + \xi_{jt} \hat{\lambda}_t^R : \quad \xi_{jt} \in [-1 \ 1] \tag{8.22}$$

By applying (8.22) on (8.21) we can rewrite the relation as:

$$z_j + \sum_{t=1}^{T} \bar{\lambda}_t^R y_{jt}^g + \sum_{t=1}^{T} \xi_{jt} \hat{\lambda}_t^R y_{jt}^g \le \sum_{t=1}^{T} U(y_{jt}) - \sum_{t=1}^{T} \sum_{i=1}^{\mathcal{N}_S} \lambda_{ijt} y_{jit} \tag{8.23}$$

Now, if the conservative mode of this relation which is called max*min* is considered, the relation (8.23) can be written as follows:

$$z_j + \sum_{t=1}^{T} \bar{\lambda}_t^R y_{jt}^g + \max_{\xi_{jt}} \sum_{t=1}^{T} \xi_{jt} \hat{\lambda}_t^R y_{jt}^g \le \sum_{t=1}^{T} U(y_{jt}) - \sum_{t=1}^{T} \sum_{i=1}^{\mathcal{N}_S} \lambda_{ijt} y_{jit} \tag{8.24}$$

The term $\max_{\xi_{jt}} \sum_{t=1}^{T} \xi_{jt} \hat{\lambda}_t^R y_{jt}^g$ should be converted into its robust counterpart form and then be utilized in the main optimization problem. Finally, the optimization problem of buyers changes to (8.25)–(8.28). The details of the procedure can be found in [32, 33].

$$\max_{y_j} \mathrm{WB}_j = z_j \\ \text{s.t.} \tag{8.25}$$

$$z_j + \sum_{t=1}^{T} \overline{\lambda}_t^R y_{jt}^g + \beta_j \Gamma_j + \sum_{t=1}^{T} \varpi_{jt} \leq \sum_{t=1}^{T} U(y_{jt}) - \sum_{t=1}^{T} \sum_{i=1}^{\mathcal{N}_S} \lambda_{ijt} y_{jit} \tag{8.26}$$

$$\varpi_{jt} + \beta_j \geq \widehat{\lambda}_t^R y_{jt}^g \tag{8.27}$$

$$\varpi_{jt} \geq 0 \tag{8.28}$$

### 8.2.6 Market Clearing Process

In the proposed market, prosumers negotiate about their desired price and amount of trading energy. First, they propose an initial price for a possible trade. Then, based on this initial price, both the buyer and seller optimize their local problem to find the optimal trading energy respecting the initial price. When they determine their desired energy, the deal price is updated based on the difference between the values of the offer and the bid of the seller and the buyer. This process is repeated until both the buyer and seller propose the same value for the price and the trading energy, which can be considered a deal. These negotiations can be modeled using iteration-based procedures.

$$\mathcal{L} = \sum_{i=1}^{\mathcal{N}_S} \mathrm{WS}_i + \sum_{j=1}^{\mathcal{N}_B} \mathrm{WB}_j + \mu g(x, y) + \lambda_{ijt}(x_{ijt} - y_{jit}) - \rho \left\| x_{ijt} - y_{jit} \right\|_2^2 \tag{8.29}$$

In this equation, $\lambda_{ijt}$ is the price parameter in the primal problem and the dual variable in the dual problem. It can be used to enable the iteration-based approach for solving decentralized optimization problems. Also, $\rho$ is a positive parameter, which is considered as the step size of the iterations.

In this case, the local problems of sellers using the augmented lagrangian method can be stated as follows:

$$\max_{x_i} \mathrm{WS}_i = \sum_{t=1}^{T} \left( \sum_{j=1}^{N_B} \lambda_{ijt} x_{ijt} - C(x_{it}) - 0.5\rho \left\| x_{ijt} - y_{jit} \right\|_2^2 \right) \\ \text{s.t.} \tag{8.30}$$

$$x_{it} = \sum_{j=1}^{N_B} x_{ijt} \tag{8.31}$$

$$x_{it}^{\min} \leq x_{it} \leq x_{it}^{\max} \tag{8.32}$$

$$C(x_{it}) = \alpha_i x_{it}^2 + \beta_i x_{it} + \gamma_i \tag{8.33}$$

$$x_{ijt} \geq 0 \tag{8.34}$$

Similarly, the optimization problem of buyers can be stated as follows:

$$\max_{y_j} \mathrm{WB}_j = z_j \tag{8.35}$$

$$\text{s.t.} \tag{8.36}$$

$$z_j + \sum_{t=1}^{T} \overline{\lambda}_t^R y_{jt}^g + \beta_j \Gamma_j + \sum_{t=1}^{T} \varpi_{jt} \leq \sum_{t=1}^{T} U(y_{jt}) - \sum_{t=1}^{T} \sum_{i=1}^{\mathcal{N}_S} \lambda_{ijt} y_{jit} - \sum_{t=1}^{T} \times \sum_{i=1}^{\mathcal{N}_S} 0.5\rho \left\| x_{ijt} - y_{jit} \right\|_2^2 \tag{8.37}$$

$$\varpi_{jt} + \beta_j \geq \widehat{\lambda}_t^R y_{jt}^g \tag{8.38}$$

$$\varpi_{jt} \geq 0 \tag{8.39}$$

Constraints (8.8)–(8.16)

At each iteration, the price, buyers' bought energy, and sellers' sold energy is calculated by the following equations:

$$x_i^{k+1} = \arg\min_x L\left(x_i, y_{jit}^{\ k}, \lambda_{ijt}^k\right) \tag{8.40}$$

$$y_j^{k+1} = \arg\min_y L\left(x_{ijt}^{\ k+1}, y_j, \lambda_{ijt}^k\right) \tag{8.41}$$

$$\lambda_{ijt}^{k+1} = \lambda_{ijt}^k - \rho\left(x_{ijt}^{\ k+1} - y_{jit}^{\ k+1}\right) \tag{8.42}$$

For accelerating the update of the step size parameter, a new approach has been proposed, which is called fast ADMM. The fast ADMM utilized in this chapter is as follows [17]:

$$\tilde{\lambda}_{ijt}^k = \lambda_{ijt}^k - \rho\left(x_{ijt}^{\ k+1} - y_{jit}^{\ k+1}\right) \tag{8.43}$$

$$\mu^{k+1} = \frac{1 + \sqrt{1 + 4(\mu^k)^2}}{2} \tag{8.44}$$

$$\alpha^{k+1} = \frac{\mu^k - 1}{\mu^{k+1}} \tag{8.45}$$

$$\lambda_{ijt}^{k+1} = \tilde{\lambda}_{ijt}^k - \alpha^{k+1}\left(\lambda_{ijt}^k - \lambda_{ijt}^{k-1}\right) \tag{8.46}$$

## 8.3 Simulation Results

In this section, the results of the developed local P2P market using the fast ADMM approach are presented and discussed. These results prove that this model can be implemented in real cases. The developed model is an optimization problem and is simulated using Python 3.6. The system used for this simulation has 6 gigabytes of RAM and an Intel Core i3 processor.

### *8.3.1 Case Study Definition*

A test system comprised of residential prosumers, including four local sellers and three local buyers is simulated in this chapter. The data of this test system is derived from [34]. The demands and production capacity of buyers and sellers are taken from [26]. The average hourly price in NYIso is considered as the price data in this model [35]. The parameters associated with the optimization problems of sellers and buyers of this test system is provided in Table 8.1.

This model will be discussed in two case studies:

- In case study 1, the developed local P2P energy trading market will be simulated with no regard to the uncertainties of the energy price in the upstream market. The nominal value for the energy price is used in this case study to solve the optimization model.
- In case study 2, the uncertainty of energy price in the upstream market is considered in the simulation process.

### *8.3.2 Case Study 1*

In this case study, the transactions between local sellers and buyers are investigated without the consideration of energy price uncertainty. The convergence speed and transactions between market players are shown in Fig. 8.1. Because this model is developed for seven prosumers that are negotiating to trade in 24 hours, the volume of obtained results is large. Therefore, we decided to provide the results for 4 hours of the day and for certain market players: 8:00 for seller #1, 12:00 for seller #2, and

**Table 8.1** Parameters of local sellers and buyers

| Producer | $\alpha_i$ (¢/kWh$^2$) | $\beta_i$ (¢/kWh) | Consumer | $\delta_j$ (¢/kWh$^2$) | $\omega_j$ (¢/kWh) |
|---|---|---|---|---|---|
| 1 | 0.0785 | 8.8287 | 1 | 0.072 | 11.4918 |
| 2 | 0.0510 | 6.9415 | 2 | 0.084 | 13.7987 |
| 3 | 0.0632 | 8.2011 | 3 | 0.045 | 10.7019 |
| 4 | 0.0614 | 5.5675 | | | |

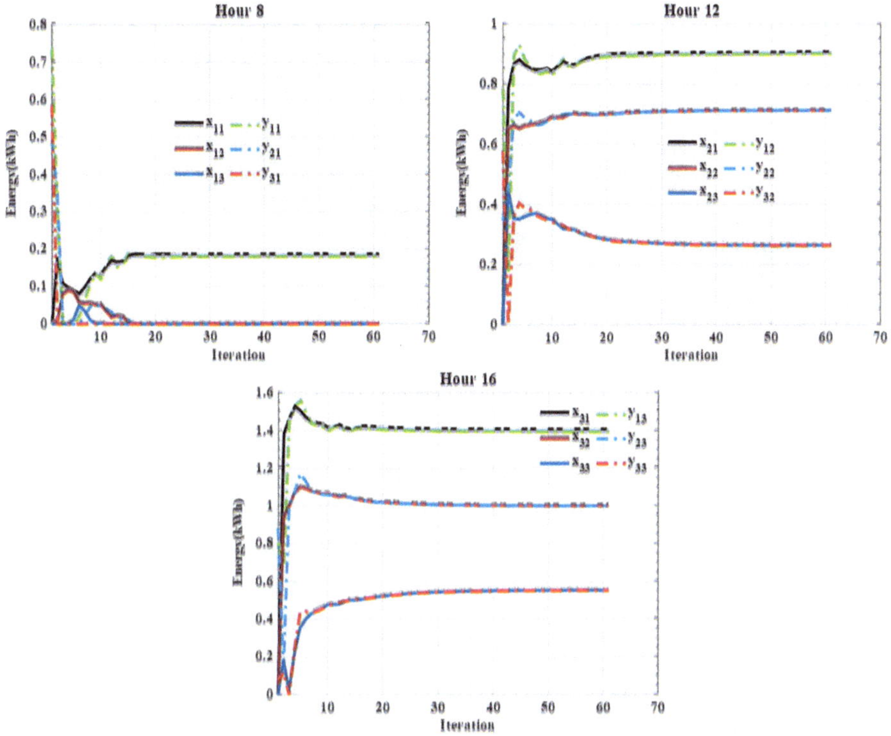


**Fig. 8.1** Energy transactions between local sellers and buyers

16:00 for seller #3. In all these subplots, the transactions of each seller with all buyers are shown. It can be observed that prosumers can make deals with multiple peers and trade a part of their desired total energy with each peer. This shows the competition, which is the most important specification of a free energy trading market. All players try to maximize their welfare in competition with other players to grab the best opportunities. Also, it can be seen from these subplots that the strategies of buyers at various hours are completely different, and they have chosen dissimilar decisions. They consider the price in the upstream market and also the average price in the local P2P energy trading market when making decisions and their energy transactions. Also, the abundance of the local energy production by sellers can alter their behavior in this market. This figure proves the convergence of the developed market and market-clearing approach. All negotiations resulted in completed deals after about 50 iterations, which is acceptable for this kind of optimization problem. This is mainly because of the novel updating method of step size that is proposed in Sect. 8.2.

As aforementioned, buyers have energy storage systems and can shift the bought energy from 1 hour to another. This helps them to reduce the cost of buying energy. The charge and discharge of the energy storage system depend on the hourly energy price. Figure 8.2 shows the charging and discharging plan of the energy storage

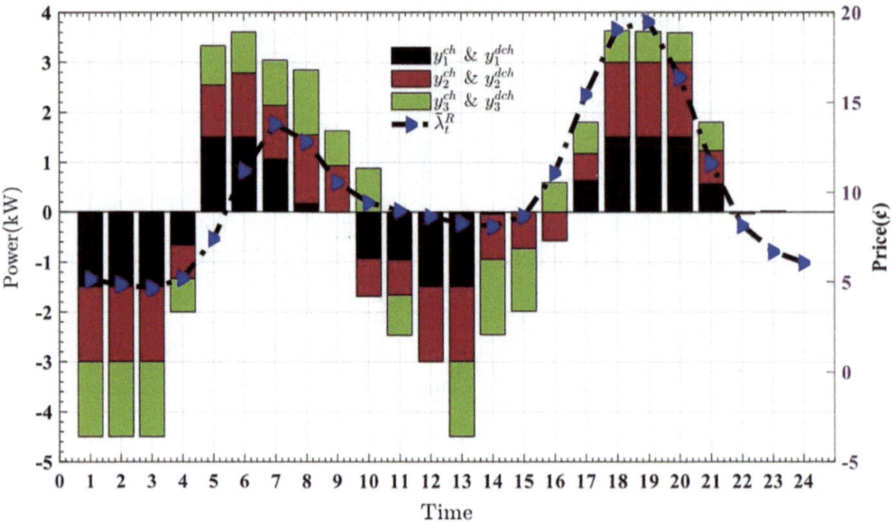


**Fig. 8.2** Charging and discharge planning of energy storage systems

systems of all buyers. The positive values show the discharge, and the negative values show the charge of these storage systems. It can be seen that during the early hours of the day when the energy price has its minimum values, all energy storage systems are charged at their maximum rate. This occurs because of the cheaper energy during these hours to shift the charged power to the hours when the demand of buyers and also the energy price are higher. These energy storage systems discharged the charged energy during the next hours, like 5–9 o'clock when the energy price witnessed a notable increase. This cycle is repeated for the next hours. The energy storage systems are charged during the mid-time of the day and are discharged at the next hours, like 17–21 o'clock, when the energy price has its maximum values. Therefore, the existence of an energy storage system can provide more flexibility for prosumers and help them to manage their transactions more wisely. Also, it is possible for prosumers to find a good opportunity in the market, but because they have consumed their desired energy, they cannot make benefit from that option. Energy storage systems can also solve this problem.

### *8.3.3 Case Study 2*

In this case study, the developed system is simulated considering the uncertainty of the energy price in the upstream market. Because the transactions and negotiations in the local P2P energy price heavily depend on the energy price in the upstream market, it can be expected that considering the uncertainty in this parameter can alter the results of the simulation.

By applying the robust approach, buyers will choose more conservative strategies in the energy markets. In other words, they consider the market price more than its nominal values to make sure that it is unlikely that their purchase cost will be more than the obtained result through the optimization process. The more they follow a conservative strategy, the less economic risk they encounter in the market. Overall, the robust approach will increase the total cost of buyers and reduce their economic risk.

In this case study, Γ is considered to be equal to 1, and the deviation of the price parameter is considered equal to 10%. This situation increased the cost of buyers for consuming electrical energy. In this case, buyers are more dependent on their own resources, including the energy storage system and the demand response, to mitigate the effects of the increase in the electrical energy price.

The transactions and convergence speed of the proposed market in case study 2, are shown in Fig. 8.3. Similar to case study 1, the volume of the output results of the cleared market is too big to show them all. Therefore, the results for 4 hours of the day and for certain market players are shown in this figure: 8:00 for seller #1, 12: 00 for seller #2, 16:00 for seller #3, and 20:00 for seller #4. It can be seen that at 12: 00, buyers have bought more energy from sellers because the local production can reduce their demand cost. They buy more energy to charge their electrical energy

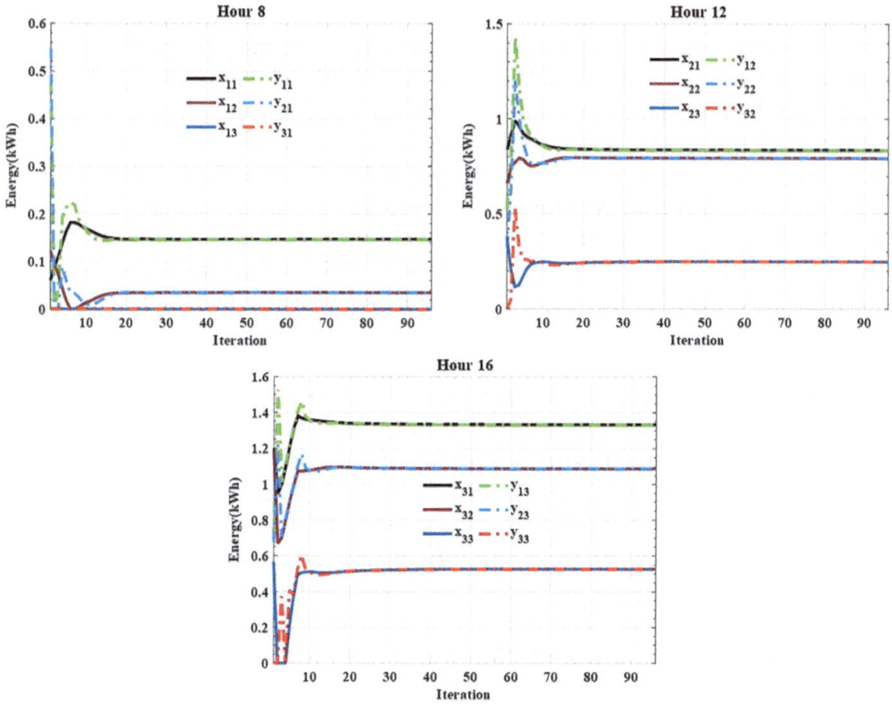


**Fig. 8.3** Energy transactions between local sellers and buyers

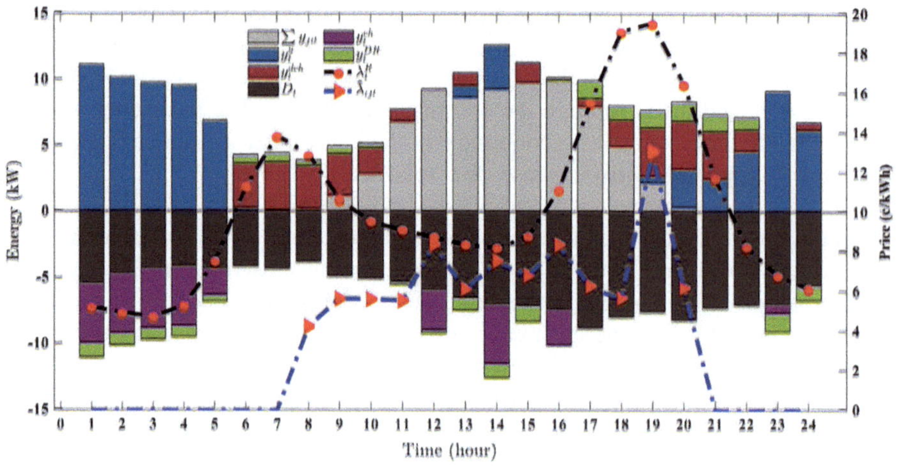


**Fig. 8.4** Power balance of buyers

storage systems and shift this charged energy to the next hours when the local production increases. The convergence of this case study is the same as in case study 1, ensuring that the market-clearing approach is performing in the right way.

When the local generation is abundant, buyers try to supply their demand by negotiating with local sellers. They can buy this energy at a lower price compared to the upstream market. As mentioned before, sellers are equipped with PVs, and they can produce energy when the solar radiation is at a level that is enough for power production by PVs. Therefore, during the mid-hours of the day, sellers can provide electrical energy for local buyers. Figure 8.4 shows the transactions between all local buyers and sellers in the P2P energy trading market. Because the energy price in the local energy market is more affordable compared to the upstream market, all buyers bought energy from local sellers, even though their demand is not so high. With some portion of this energy, they can charge their energy storage systems and consume the charged energy during the next hours, when the local sellers are unable to generate energy.

Figure 8.5 shows the power balance of buyers in the local energy trading market. This figure depicts all resources that buyers can control, including energy transactions in the market, energy storage systems, and the demand response. It can be seen that buyers have tried to use all capacity of energy storage systems and demand response to shift their energy transactions to the early hours of the day. Then, they use this shifted energy during other hours, to reduce their cost. For example, it can be seen that during hours 6–10, buyers have discharged a portion of the stored energy in the energy storage systems and reduced a percent of their hourly demand. During this time, the energy price has seen one of its peaks. Therefore, buyers attempt to reduce their energy transactions with energy markets during this period. During the next hours, from 11 to 16 o'clock, the performance of energy storage systems and demand response have been reversed. During this period, buyers have increased

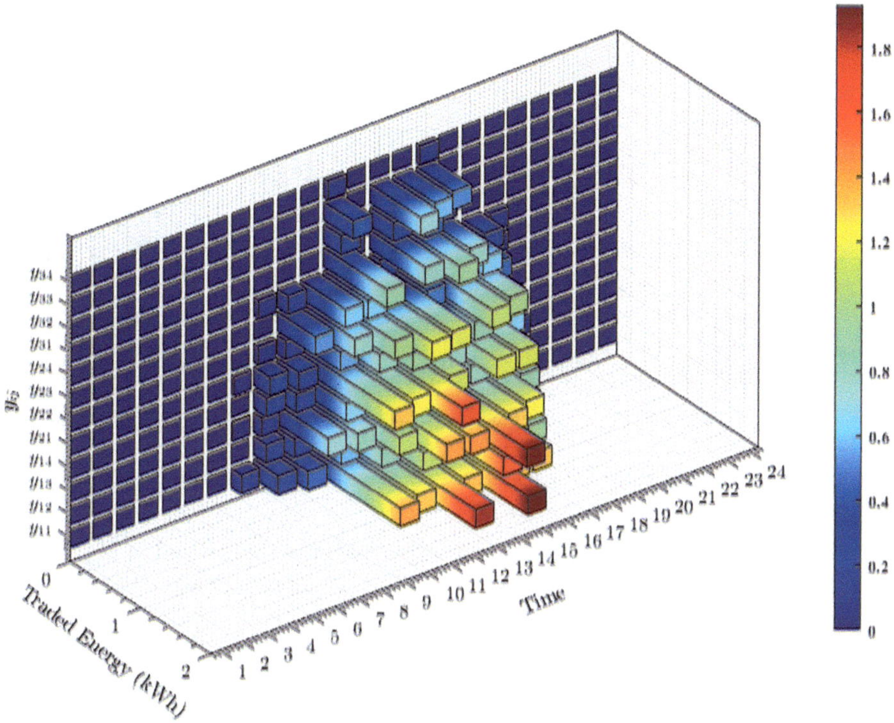


**Fig. 8.5** Energy transactions between all local buyers and sellers

**Table 8.2** Comparison of players' welfare with and without considering the price uncertainty

| Welfare (¢) | $WS_1$ | $WS_2$ | $WS_3$ | $WS_4$ | $WB_1$ | $WB_2$ | $WB_3$ | Total |
|---|---|---|---|---|---|---|---|---|
| With robust | 88.8226 | 35.1784 | 9.6828 | 4.8914 | −2.123 | 183.1816 | 156.0819 | 475.7157 |
| No robust | 88.602 | 35.173 | 9.645 | 4.802 | 2.1267 | 184.665 | 157.42 | 482.4337 |

their demand by charging the energy storage systems and also shifting a part of their hourly demand related to other hours to these hours. This cycle repeated for the next hours of the day. It can be understood that all activities of prosumers are dependent on the electrical energy price. This is a result of prosumers being profit-based agents.

Considering the uncertainty of the energy price can affect the welfare of market players. The robust approach increases the energy price, which can increase the welfare of sellers. On the other hand, an increase in the price can reduce the welfare of buyers. Table 8.2 states the welfare of market players in case studies 1 and 2. It can be seen that the robust approach has increased the welfare of sellers and decreased the welfare of buyers.

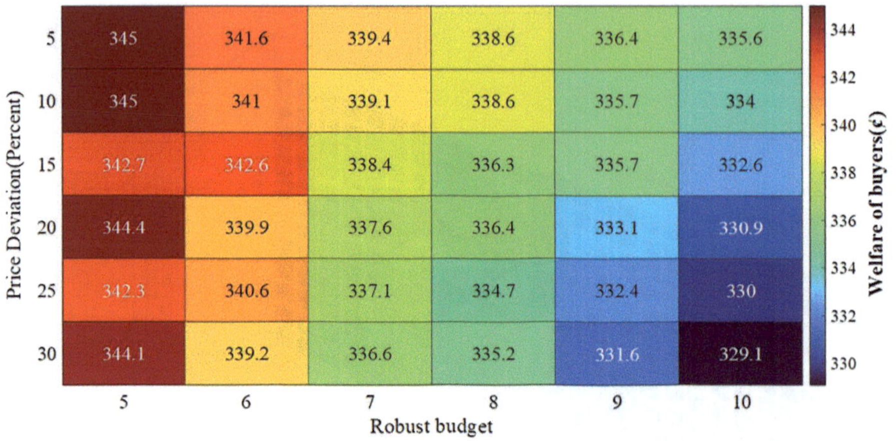


**Fig. 8.6** Effects of price deviation and budgeting coefficient on the welfare of players

The effects of the robust approach are investigated in more detail in Fig. 8.6. In this figure, the alteration of players because of the changes in budgeting coefficient and price deviation is shown. It can be seen that when the Γ increases, the market players reach more total welfare. In contrast, any growth in the price deviation decreases the total welfare of market players. The price variation demonstrates the robustness of the problem. Thus, when the players follow a conservative strategy, their welfare witnesses a decrease in its value.

## 8.4 Conclusion

In this chapter, a fully decentralized P2P energy trading market for small-scale prosumers, including local sellers and buyers, was developed and presented. The prosumers in this chapter are considered economic-based agents who are seeking the maximum welfare resulting from participating in the local energy trading market. In this market, all prosumers negotiate with each other to find the best peers. This is a fully decentralized market in which market players can make decisions based on their own priorities and freely trade with other players. Also, in the developed model, the privacy of players is secured, and there is no supervision from any third parties. The results proved the effectiveness and the convergence of the proposed market-clearing approach. The fast ADMM approach was used in this chapter to clear the developed local market. This method is strengthened with a novel method of step size updating, which resulted in a notable decrease in iteration needed for the convergence. Moreover, the uncertainty of electrical energy price was considered in this chapter. The robust optimization model is utilized to deal with this uncertainty. In the simulation results section, we investigated the impacts of the robustness on the market players' behavior. Overall, it was observed that energy transactions in the local P2P energy trading market are related to the energy price and the robustness parameters value. The results showed that by increasing the robustness, the total welfare of market players decreases. Also, robustness affected the scheduling of energy storage systems and the demand response program.

## References


1. Gottwalt, S., Gärttner, J., Schmeck, H., & Weinhardt, C. (2016). Modeling and valuation of residential demand flexibility for renewable energy integration. *IEEE Transactions on Smart Grid, 8*, 2565–2574.
2. Tushar, W., Saha, T. K., Yuen, C., Azim, M. I., Morstyn, T., Poor, H. V., et al. (2020). A coalition formation game framework for peer-to-peer energy trading. *Applied Energy, 261*, 114436. https://doi.org/10.1016/j.apenergy.2019.114436
3. Xiao, Y., Wang, X., Pinson, P., & Wang, X. (2018). A local energy market for electricity and hydrogen. *IEEE Transactions on Power Apparatus and Systems, 33*, 3898–3908. https://doi.org/10.1109/TPWRS.2017.2779540
4. Xiao, Y., Xiao, Y., Wang, X., Pinson, P., & Wang, X. (2020). Transactive energy based aggregation of prosumers as a retailer. *IEEE Transactions on Smart Grid, 11*, 3302–3312. https://doi.org/10.1109/TSG.2020.2976130
5. Nojavan, S., Zare, K., & Mohammadi-Ivatloo, B. (2017). Optimal stochastic energy management of retailer based on selling price determination under smart grid environment in the presence of demand response program. *Applied Energy, 187*, 449–464. https://doi.org/10.1016/j.apenergy.2016.11.024
6. Zhang, C., Wu, J., Cheng, M., Zhou, Y., & Long, C. (2016). A bidding system for peer-to-peer energy trading in a grid-connected microgrid. *Energy Procedia, 103*, 147–152. https://doi.org/10.1016/j.egypro.2016.11.264
7. Devine, M. T., & Cuffe, P. (2019). Blockchain electricity trading under demurrage. *IEEE Transactions on Smart Grid, 10*, 2323–2325. https://doi.org/10.1109/TSG.2019.2892554
8. Zhu, M., & Martinez, S. (2012). On distributed convex optimization under inequality and equality constraints. *IEEE Transactions on Automatic Control, 57*, 151–164. https://doi.org/10.1109/TAC.2011.2167817
9. Zhang, H., Shi, W., Mokhtari, A., Ribeiro, A., & Ling, Q. (2017). Decentralized constrained consensus optimization with primal dual splitting projection. In *2016 IEEE global conference on signal and information processing (GlobalSIP) 2016 – Proceedings* (pp. 565–569). https://doi.org/10.1109/GlobalSIP.2016.7905905
10. Khorasany, M., Mishra, Y., & Ledwich, G. (2020). A decentralized bilateral energy trading system for peer-to-peer electricity markets. *IEEE Transactions on Industrial Electronics, 67*, 4646–4657. https://doi.org/10.1109/TIE.2019.2931229
11. Baroche, T., Pinson, P., Latimier, R. L. G., & Ben Ahmed, H. (2019). Exogenous cost allocation in peer-to-peer electricity markets. *IEEE Transactions on Power Apparatus and Systems, 34*, 2553–2564. https://doi.org/10.1109/TPWRS.2019.2896654
12. Morstyn, T., & McCulloch, M. D. (2018). Multiclass energy management for peer-to-peer energy trading driven by prosumer preferences. *IEEE Transactions on Power Systems, 34*(5), 4005–4014.
13. Boyd, S., Parikh, N., Chu, E., Peleato, B., & Eckstein, J. (2011). *Distributed optimization and statistical learning via the alternating direction method of multipliers*. Now Publishers Inc.
14. Munsing, E., Mather, J., & Moura, S. (2017, January). Blockchains for decentralized optimization of energy resources in microgrid networks. In *1st Annual IEEE conference on control technology and applications, CCTA 2017* (pp. 2164–2171). https://doi.org/10.1109/CCTA.2017.8062773

15. Moret, F., & Pinson, P. (2019). Energy collectives: A community and fairness based approach to future electricity markets. *IEEE Transactions on Power Apparatus and Systems, 34*, 3994–4004. https://doi.org/10.1109/TPWRS.2018.2808961
16. Lin, W.-T., Chen, G., & Li, C. (2021). Risk-averse energy trading among peer-to-peer based virtual power plants: A stochastic game approach. *International Journal of Electrical Power & Energy Systems, 132*, 107145. https://doi.org/10.1016/j.ijepes.2021.107145
17. Ullah, M. H., & Park, J. D. (2021). Peer-to-peer energy trading in transactive markets considering physical network constraints. *IEEE Transactions on Smart Grid*, 3053. https://doi.org/10.1109/TSG.2021.3063960
18. Sorin, E., Bobo, L., & Pinson, P. (2019). Consensus-based approach to peer-to-peer electricity markets with product differentiation. *IEEE Transactions on Power Apparatus and Systems, 34*, 994–1004. https://doi.org/10.1109/TPWRS.2018.2872880
19. Hug, G., Kar, S., & Wu, C. (2015). Consensus + innovations approach for distributed multiagent coordination in a microgrid. *IEEE Transactions on Smart Grid, 6*, 1893–1903. https://doi.org/10.1109/TSG.2015.2409053
20. Esmat, A., de Vos, M., Ghiassi-Farrokhfal, Y., Palensky, P., & Epema, D. (2021). A novel decentralized platform for peer-to-peer energy trading market with blockchain technology. *Applied Energy, 282*, 116123. https://doi.org/10.1016/j.apenergy.2020.116123
21. Paudel, A., Sampath, L. P. M. I., Yang, J., & Gooi, H. B. (2020). Peer-to-peer energy trading in smart grid considering power losses and network fees. *IEEE Transactions on Smart Grid, 11*, 4727–4737. https://doi.org/10.1109/TSG.2020.2997956
22. Chernov, A., Dvurechensky, P., & Gasnikov, A. (2016). Fast primal-dual gradient method for strongly convex minimization problems with linear constraints. In *International conference on discrete optimization and operations research* (pp. 391–403). Springer.
23. Morstyn, T., Teytelboym, A., & McCulloch, M. D. (2019). Bilateral contract networks for peer-to-peer energy trading. *IEEE Transactions on Smart Grid, 10*, 2026–2035. https://doi.org/10.1109/TSG.2017.2786668
24. Mehdinejad, M., Shayanfar, H., & Mohammadi-Ivatloo, B. (2022). Peer-to-peer decentralized energy trading framework for retailers and prosumers. *Applied Energy, 308*. https://doi.org/10.1016/j.apenergy.2021.118310
25. Elkazaz, M., Sumner, M., & Thomas, D. (2021). A hierarchical and decentralized energy management system for peer-to-peer energy trading. *Applied Energy, 291*. https://doi.org/10.1016/j.apenergy.2021.116766
26. Mehdinejad, M., Shayanfar, H., & Mohammadi-Ivatloo, B. (2022). Decentralized blockchain-based peer-to-peer energy-backed token trading for active prosumers. *Energy, 244*, 122713. https://doi.org/10.1016/j.energy.2021.122713
27. Cioara, T., Antal, M., Mihailescu, V. T., Antal, C. D., Anghel, I. M., & Mitrea, D. (2021). Blockchain-based decentralized virtual power plants of small prosumers. *IEEE Access, 9*, 29490–29504. https://doi.org/10.1109/ACCESS.2021.3059106
28. Guerrero, J., Sok, B., Chapman, A. C., & Verbič, G. (2021). Electrical-distance driven peer-to-peer energy trading in a low-voltage network. *Applied Energy, 287*, 116598. https://doi.org/10.1016/j.apenergy.2021.116598
29. Muyeen, S. M. (2020). Peer-to-peer energy trading in virtual power plant based on blockchain smart contracts. *IEEE Access, 8*, 175713–175726. https://doi.org/10.1109/ACCESS.2020.3026180
30. Jogunola, O., Wang, W., & Adebisi, B. (2020). Prosumers matching and least-cost energy path optimisation for peer-to-peer energy trading. *IEEE Access, 8*, 95266–95277.
31. Samadi, P., Mohsenian-Rad, H., Schober, R., & Wong, V. W. S. (2012). Advanced demand side management for the future smart grid using mechanism design. *IEEE Transactions on Smart Grid*. https://doi.org/10.1109/TSG.2012.2203341

32. Li, Z., Ding, R., & Floudas, C. A. (2011). A comparative theoretical and computational study on robust counterpart optimization: I. Robust linear optimization and robust mixed integer linear optimization. *Industrial & Engineering Chemistry Research*. https://doi.org/10.1021/ie200150p
33. Bertsimas, D., & Sim, M. (2004). The price of robustness. *Operations Research*. https://doi.org/10.1287/opre.1030.0065
34. Dataport | Login | Signup. (n.d.)
35. Historical Data | ENGIE Resources | Commercial Energy Provider. (n.d.)

# Chapter 9
# Peer-to-Peer Energy Trading in Multi-carrier Energy Systems

**Mohammad Hasan Ghodusinejad and Hossein Yousefi**

## 9.1 Energy Market Transformation

The structure of the energy market has changed during the last decade in various aspects. Due to the expansion of the use of distributed renewable energy sources (RES) and the emergence of new energy generation and storage technologies, along with the increase in energy consumption and the emergence of new energy consumption patterns such as electric vehicles, this structural evolution on the consumer side has gained a lot of momentum. In general, it can be said that numerous changes in energy networks have pursued the goal of involving energy consumers in energy management through different mechanisms. Over the past years and in different countries, in energy networks undergoing restructuring, consumers who have become energy producers due to distributed energy sources, are integrated with the energy network and exchange their excess energy through certain policies [1].

According to the introduction of demand-side management mechanisms, the main approach to restructuring energy networks has been to increase the participation of final energy consumers in the energy management process. In general, this process can be summarized in three main stages. Figure 9.1 depicts the general structure of these three stages. As seen in Fig. 9.1, the structure of the energy network has altered from fully unidirectional to fully bidirectional. According to the energy exchange mechanisms in these networks, they can be named and described as follows:

M. H. Ghodusinejad · H. Yousefi (✉)
Renewable Energies and Environmental Engineering Department, Faculty of New Sciences and Technologies, University of Tehran, Tehran, Iran
e-mail: hosseinyousefi@ut.ac.ir



V. Vahidinasab, B. Mohammadi-Ivatloo (eds.), *Demand-Side Peer-to-Peer Energy Trading*, Green Energy and Technology, https://doi.org/10.1007/978-3-031-35233-1_9

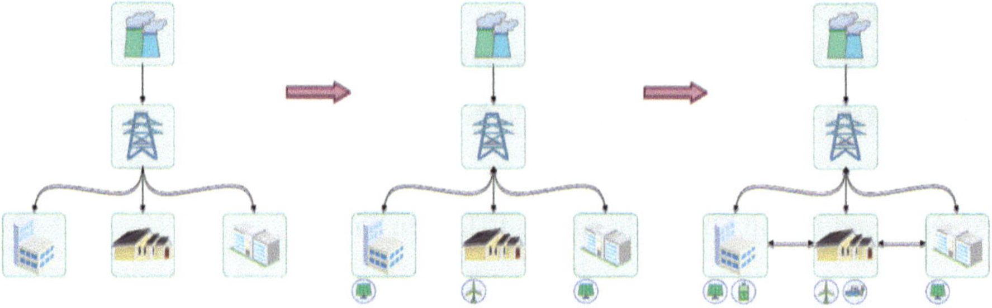

**Fig. 9.1** Three-stage evolution of energy networks

- *Conventional Network*: Traditional energy networks operate in a completely unidirectional way and rely especially on bulk and centralized energy production. In these networks, energy is produced in centralized power plants (mainly fossil fuel) and reaches the final consumer through energy transmission lines. In conventional networks, centralized energy management is totally applied and consumers are not involved in the energy market [2].
- *Semi-transactive Network*: The boom in the use of distributed energy resources (DERs), which are mainly based on renewable energies, led to the creation of semi-transactive markets; so that consumers have become energy producers by using DERs and are able to transfer energy to the grid. Various mechanisms, such as net metering, feed-in-tariff (FiT), etc., have been introduced to the energy markets at this stage.
- *Full-transactive Network*: With the transition from semi-transactive to full-transactive networks, the framework of energy exchange in energy networks has undergone a transformation. In semi-transactive networks, final consumers have only been allowed to exchange energy with the main grid, that is, retailers and energy distribution companies, while in the new generation of energy networks, energy exchange is not only conducted with the main grid but also between consumers. In this context, the *peer-to-peer energy sharing* concept is introduced. In fact, today, one of the most promising solutions for the restructuring of energy networks is the peer-to-peer (P2P) energy exchange mechanism. The emergence of more active consumers, or so-called "producers/consumers" or "prosumers" with energy production and storage capabilities, empowers them in order to provide new opportunities for the performance of energy systems. Peer-to-peer (P2P) electricity markets conceptually allow consumers to share electrical energy directly. These P2P markets rely on a bottom-up, consumer-centric approach by giving consumers the opportunity to freely choose how to purchase the electricity they need [3].

## 9.2 P2P Energy Trading

The core of peer-to-peer energy trading is based on the fundamental concept of "prosumers". In fact, the term "peer" in such energy networks indicates the parties connected to the network, which exist in two forms: regular consumers and prosumers. In general, peer-to-peer (P2P) energy trading allows all peers to actively participate in the energy market and earn various benefits based on incentives [1].

Peers or prosumers are the final links in the energy supply chain. The process of transition and evolution in electricity markets, during the last two decades, has generally been based on increasing the participation of these final links in market management. By going through this process, end consumers will become active members of the energy market rather than passive members. The beginning of this process commenced with the introduction of demand response (DR) and demand side management (DSM) programs and continued with peer-to-grid (P2G) energy exchange. Today, with the emergence of peer-to-peer energy trading, along with other concepts of the past, the intervention of end users of the energy chain in energy management has significantly increased. This process is briefly and conceptually shown in Fig. 9.2.

Peer-to-peer energy trading is a business model in which market design particularly plays an important role. Accordingly, P2P-based markets are divided into three general categories: centralized (community-based market), decentralized (full P2P market), and distributed (hybrid P2P market) [3, 4]:

- *Centralized (community-based market)*: In a centralized marketplace, there is a coordinator or community manager that communicates with each peer. Based on the information gathered from the peers, the coordinator directly decides on the energy exchange of the peers. Obviously, a community is a collection of members that share common interests and goals, such as maximizing total income. Regarding some predefined rules, the community manager distributes the income of the entire P2P community. Hence, one of the major advantages of a community-based market is that it can maximize the social welfare of the entire market. Reducing the authority of peers and jeopardizing the privacy of market members are among the disadvantages of these markets.

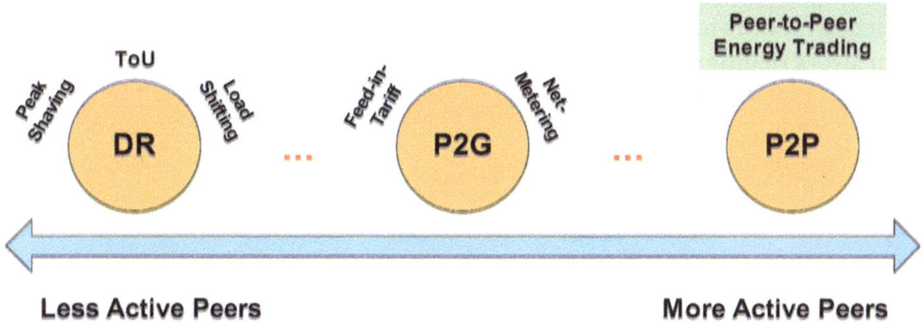


**Fig. 9.2** Contribution trend of peers in market energy management

- *Decentralized (full P2P market)*: This type of market is based on negotiation and direct energy exchange between peers. Therefore, unlike centralized markets, there is no market manager in decentralized P2P energy trading markets. One of the main advantages of these markets is that the privacy of peers is well protected. In addition, the scalability of decentralized markets is better and peers can easily join and leave the market. In contrast, the dynamic nature and lack of a coordinator can result in the output of the decentralized P2P market not being very lucid or predictable to the network.
- *Distributed (hybrid P2P market)*: Distributed markets are between centralized and decentralized markets in terms of design; As a result, it can be said that they simultaneously benefit from the advantages of the two previous types and minimize their disadvantages. For this reason, these markets are called hybrid P2P markets. In hybrid markets, although there is a coordinator, it does not directly affect the exchange of energy between peers and usually affects the market trend by sending pricing signals, etc. Compared to decentralized markets, better control and coordination of peers' behaviors can be seen in distributed markets due to the presence of a coordinator. Also, compared to centralized markets, distributed markets typically provide a higher level of privacy and independence for peers.

## 9.3 Multi-energy P2P Systems

### *9.3.1 Conceptual Analysis*

Due to the close relationship between peer-to-peer energy trading and electricity market structures, research and project implementation in this area have often focused on the electricity exchange. The sharing economy, which is one of the basic principles of peer-to-peer energy trading, is usually more implementable in the field of electric energy; Because the exchange of electrical energy in the world has relatively more ready infrastructure compared to other energy carriers. However, in some research, other energy carriers have opened up to the topic of P2P energy trading. As a result, the topic of *multi-energy P2P systems* has been introduced and taken into consideration.

In the past, the networks of different energy carriers were designed and developed separately. Meanwhile, with the emergence of a systematic view of the energy field, the network of different energy carriers became intertwined. With increasing awareness regarding the benefits of using multi-energy systems (MES) and, at the same time, expanding the generation and storage technologies of different energy carriers and related distribution infrastructures, the acceptance of these systems increased dramatically. This greatly contributed to the development of the energy hub (EH) concept. A common definition of an energy hub is as follows: a place where the production, conversion, storage, and consumption of different energy carriers take place [5].

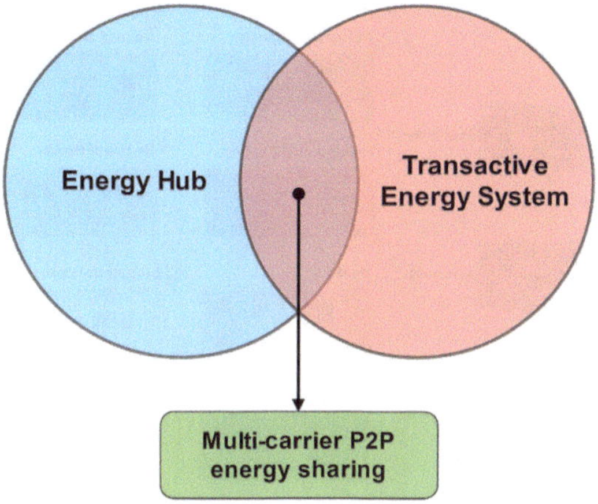


**Fig. 9.3** Conceptual structure of multi-energy P2P system compared to energy hub and transactive energy system

On the other hand, transactive energy systems (TES) have been introduced as another promising solution to overcome the challenges of energy networks. Transactive energy systems are a set of economic, control and technological mechanisms that allow microgrids to optimize the production and consumption of energy at the same time and provide the possibility of energy sharing among members. This leads to optimal management of the energy production/consumption pattern and ultimately increases the technical and economic efficiency of energy systems.

The overlay of the two basic concepts of energy hub and transactive energy system leads to the introduction of a multi-energy P2P system. In fact, from a technical and economic point of view, the multi-energy P2P system is the intersection of the two above-mentioned concepts. Figure 9.3 illustrates this issue.

Multi-energy P2P systems can bring advantages and positive characteristics of energy hubs and transactive energy systems at the same time. These systems are able to create a dynamic balance between supply and demand by providing energy exchange opportunities at different levels. At the same time, with the possibility of using components of production, conversion, and storage of different energy carriers, they adopt an integrated approach to the energy system and lead to an increase in the technical and economic efficiency of the energy system.

### *9.3.2 System Components*

In a multi-energy P2P system, various devices for producing, converting, and storing different energy carriers can be integrated. In addition, others, such as non-fossil vehicles or water-related systems, are included in this category. Figure 9.4 shows a general view of the components in a multi-energy P2P system.

Distributed energy generation systems mainly include renewable sources (wind and solar) and non-renewable sources (gas or diesel generators). Some components

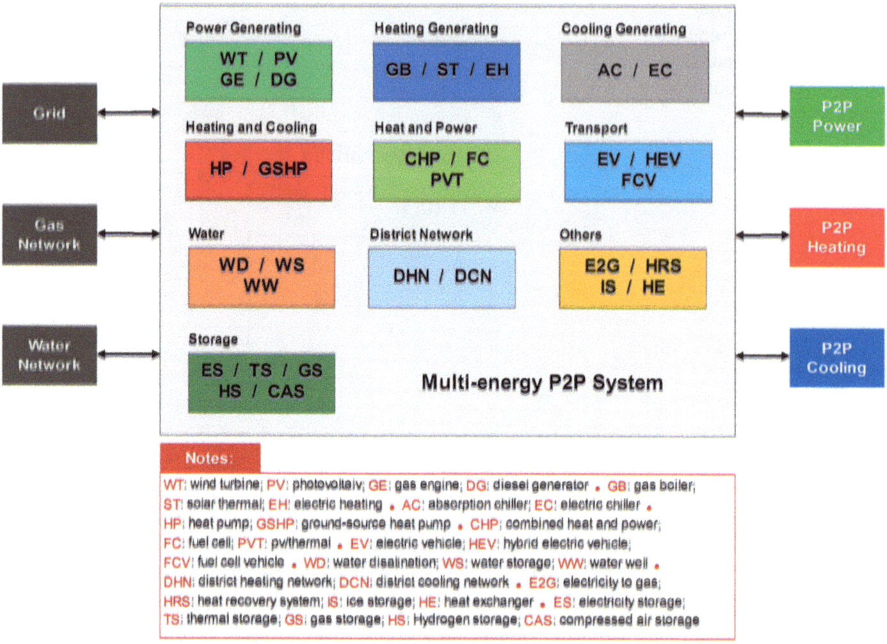


**Fig. 9.4** Common components of the multi-energy P2P system

specifically produce cooling or heating, and some can produce both. In addition, heat and power cogeneration systems are prominent and inseparable components of multi-energy P2P systems.

One of the important components to be mentioned is energy storage systems. Depending on the different energy carriers used in the system, different types of storage devices can be integrated. The most common energy storage is batteries, while other types of energy storage technologies are also used.

### *9.3.3 A Brief Literature Review*

Although most of the research in the field of P2P energy trading has focused on electricity-based systems, a number of papers have worked on multi-energy P2P systems. In these papers, in addition to electricity, the sharing of other carriers such as heating, cooling, gas, and even water is emphasized. In some articles, only the electricity trading is conducted and other carriers are only coupled among the members, while in others, all energy carriers are traded.

In recent research, Wang et al. [6] proposed a multi-energy P2P market in order to allow members to trade both electricity and heat. Two different trading scenarios, namely an electricity-only trading coalition and an electricity–heat trading coalition, were predefined. The peers were designed to join one of these two coalitions according to their potential benefits and trade energy inside the coalition. In another work, Zhong et al. [7] focused on an optimal energy management strategy to minimize the operation cost of a multi-energy P2P system, considering technical and environmental constraints. The proposed energy management strategy was comprised of a day-ahead phase and an intra-day phase to overcome the uncertain effects of power generation from renewable energy sources as well as load demands.

Jalilian et al. [8] suggested an integrated scheduling model for optimal dispatch of cooling, heating, power, gas, and water sources in an energy-water microgrid, where the microgrid operator participates in the power, heat, and gas markets and utilizes energy conversion facilities to meet various demands. In this context, a multi-objective, two-stage stochastic optimization problem is presented to minimize the total operating and emission costs as well as the amount of potable water extracted from water wells. Javadi et al. [9] also investigated a fully decentralized model for energy trading within a multi-energy P2P market. The optimal operation of the energy hubs was modeled as a standard mixed-integer linear programming (MILP) optimization problem.

The research conducted in relation to multi-energy P2P systems can be examined from various aspects, such as different components and equipment used in the system, various energy carriers, etc. Table 9.1 provides details of recent papers in this field.

**Table 9.1** Details of recent research regarding multi-energy P2P systems

| Author | Publish year | System components | Carriers/ flows | Objective function |
|---|---|---|---|---|
| Wang et al. [6] | 2022 | PV/WT/HP | Electricity/ heating | Maximize the net benefit |
| Zhong et al. [7] | 2022 | WT/ES/HP/ CHP/GB/TS | Electricity/ heating/gas | Minimize the social operation cost |
| Jalilian et al. [8] | 2022 | WT/EH/GB/ CHP/ES/T/ GS/IS/AC/WD/ WS/WW | Electricity/ cooling/ heating/gas/ water | Minimize the total cost and amount of potable water extracted from water wells |
| Javadi et al. [9] | 2022 | PV/ES/EH/HP/ CHP/GB/AC | Electricity/ cooling/ heating/gas | Minimize the total cost |
| Zhang et al. [10] | 2022 | PV/ES/CHP/HP | Electricity/ cooling/ heating/gas | Maximize the net benefit |
| Daneshvar et al. [11] | 2022 | WT/PV/CCHP/ ES/TS | Electricity/ cooling/ heating/gas | Minimize operating costs |
| Zhang et al. [12] | 2022 | WT/PV/ES/WD/ WS/GS/GE/E2G | Electricity/ gas/water | Minimize operating costs |
| Daneshvar et al. [13] | 2022 | PV/GE/GB/AC/ ES/TS | Electricity/ cooling/ heating | Minimize operating costs |

(continued)

**Table 9.1** (continued)

| Author | Publish year | System components | Carriers/ flows | Objective function |
|---|---|---|---|---|
| Davoudi and Moeini-Aghtaie [14] | 2021 | PV/ES/EC/CHP/ GB/TS/ST | Electricity/ cooling/ heating/gas | Maximize welfare |
| Liu et al. [15] | 2021 | WT/PV/ES/FC/ HS/HE | Electricity/ hydrogen | Maximize self-consumption, minimize carbon emission and utility bills |
| Davoudi et al. [16] | 2021 | PV/ES/EC/CHP/ GB/TS/ST | Electricity/ cooling/ heating/gas | Maximize the net profit |
| Khorasany et al. [17] | 2021 | WT/PV/GE/GB/ ST/ES/EC/TS/ AC/IS | Electricity/ cooling/ heating/gas | Minimize operating costs |
| Li and Zhang [18] | 2021 | PV/CHP | Electricity/ heating/gas | Minimize operating costs |
| Nguyen and Ishihara [19] | 2021 | FC/HE/TS/GB/ HRS/CAS | Electricity/ heating/gas | Minimize energy costs |
| Daneshvar et al. [20] | 2020 | WT/PV/TS/ES/ EC/AC | Electricity/ cooling/ heating | Maximize the cost-savings |
| Liu et al. [21] | 2018 | PVT/HP/CHP | Electricity/ heating/gas | Minimize operating costs |

## 9.4 Illustrative Case Study

In order to investigate more precisely the performance of multi-energy P2P systems, a case study example has been presented. In this example, a multi-energy P2P system has been studied in the presence of combined heat and power sources and in the framework of an optimization model.

### *9.4.1 System Design*

Figure 9.5 shows the structure of each peer and its components. As can be seen, the components of each member include PVT panels, an internal combustion engine (ICE), a gas boiler, an electric chiller (EC), and an absorption chiller (AC). Each peer has a specific demand for power, cooling, and heating, depending on its application. The power requirement is provided through PVT, ICE, the power grid and other peers. The heating of each peer is provided through internal sources, i.e., PVT, ICE, and boiler, as well as heating coupling with other peers, while the required cooling is provided by internal absorption and electric chillers and imports from other peers.

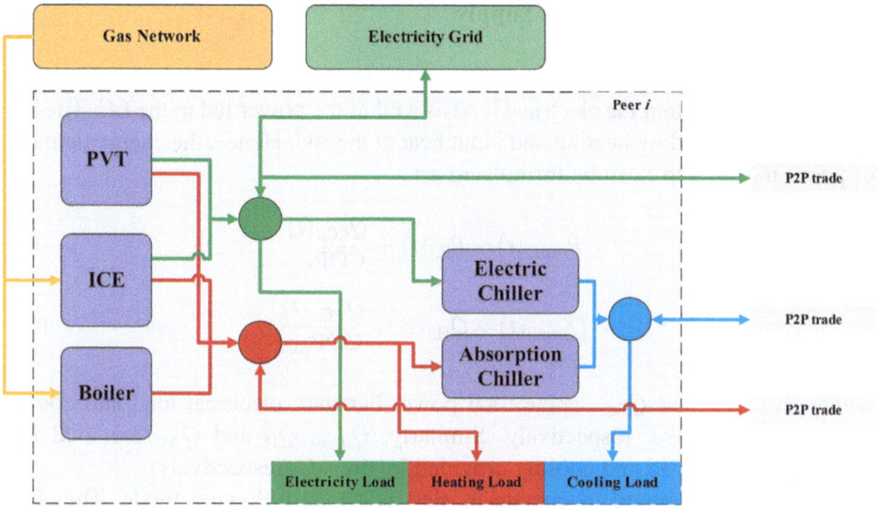


**Fig. 9.5** The structure of one single peer in the market

**Table 9.2** Technical characteristics of the system components [22]

| | |
|---|---|
| PV/T reference electrical efficiency (%) | 15.5 |
| Temperature coefficient (%/°C) | 0.43 |
| Heat loss coefficient (W/m$^2$.K) | 9.12 |
| ICE electrical efficiency coefficients (%) | $a = 0.1$, $b = 0.4$, $c = -0.2$ |
| Heat recovery system efficiency (%) | 80 |
| Boiler efficiency (%) | 80 |
| COP of absorption chiller ($Wh_{th}/Wh_{th}$) | 0.7 |
| COP of electric chiller ($Wh_{th}/Wh_{el}$) | 3 |

The studied system is designed in the form of an optimization problem; so that the decision variables include the optimal sizes of components for each peer in order to minimize the operating costs. The decision variables include the number of PVT panels, ICE capacity, and AC capacity. The capacity of the boiler and EC is also determined depending on the optimal values of the variables and the internal energy needs of each member. This optimal sizing problem is designed in the context of a P2P market and is solved using a genetic algorithm. For simplicity, the problem is implemented for a market with two members (a commercial building (as peer 1) and an office building (as peer 2)). The technical characteristics of the system components are presented in Table 9.2.

#### 9.4.1.1 Energy Demand and Supply

The total energy demand of each peer is in the form of heat and power. The power demand is for meeting the electrical load as well as the power fed to the EC. The heat is required for building heating and input heat of the AC. Hence, the energy demands of peer $i$ in timestep $t$ can be formulated as:

$$P_{\mathrm{dem},i}(t) = P_{\mathrm{L},i}(t) + \frac{Q_{\mathrm{EC},i}(t)}{\mathrm{COP_{EC}}} \tag{9.1}$$

$$Q_{\mathrm{dem},i}(t) = Q_{\mathrm{H},i}(t) + \frac{Q_{\mathrm{AC},i}(t)}{\mathrm{COP_{AC}}} \tag{9.2}$$

where $P_{\mathrm{dem}}$, $P_{\mathrm{L}}$ and $Q_{\mathrm{EC}}$ denote total power demand, electrical load and cooling provided by the EC, respectively. Similarly, $Q_{\mathrm{dem}}$, $Q_{\mathrm{H}}$ and $Q_{\mathrm{AC}}$ are total heat demand, heating load and cooling provided by the AC, respectively.

The ICE is supposed to operate in electrical load-following mode. Therefore, since the electricity demand in some hours is less than the nominal capacity of the ICE, it has to operate in partial load mode and so that the electrical efficiency would be reduced. Hereby, the partial load ratio is defined as follows:

$$\mathrm{PLR}_{\mathrm{ICE},i}(t) = \frac{P_{\mathrm{dem},i}(t)}{P_{\mathrm{ICE-nom},i}} \leq 1 \tag{9.3}$$

where $\mathrm{PLR_{ICE}}$ and $P_{\mathrm{ICE-nom}}$ are partial load ratio and nominal capacity of the ICE.

Then, the ICE efficiency is a function of the partial load ratio and is determined by Eq. (9.4). The output power of the ICE is also calculated by (9.5) [22]:

$$\eta_{\mathrm{ICE},i}(t) = a + b.\mathrm{PLR}_{\mathrm{ICE},i}(t) + c.\mathrm{PLR}^2_{\mathrm{ICE},i}(t) \tag{9.4}$$

$$P_{\mathrm{ICE},i}(t) = \begin{cases} 0 & 0 \leq \mathrm{PLR}_{\mathrm{ICE},i}(t) < 25\% \\ P_{\mathrm{dem},i}(t) & 25\% \leq \mathrm{PLR}_{\mathrm{ICE},i}(t) < 100\% \\ P_{\mathrm{ICE-nom},i} & \mathrm{PLR}_{\mathrm{ICE},i}(t) = 100\% \end{cases} \tag{9.5}$$

The ICE fuel consumption is a function of its output power and efficiency and can be found as follows:

$$Q_{\mathrm{f-ICE},i}(t) = \frac{P_{\mathrm{ICE},i}(t)}{\eta_{\mathrm{ICE},i}(t)} \tag{9.6}$$

Finally, the generated heat by ICE is a function of the ICE and heat recovery system efficiencies. Hence, it can be found by (9.7):

$$Q_{\mathrm{ICE},i}(t) = \eta_{\mathrm{HRS}} Q_{\mathrm{f-ICE},i}(t)\left(1 - \eta_{\mathrm{ICE},i}(t)\right) \tag{9.7}$$

where $\eta_{\mathrm{HRS}}$ and $\eta_{\mathrm{ICE}}$ are the efficiencies of heat recovery system and ICE, respectively.

For the PV/T panels considered in this paper, the area of each panel is about 1.6 m$^2$. Therefore, the total area of the PV/T system for each peer, namely $A_{\mathrm{PVT},\ i}$, can be calculated as $1.6N_{\mathrm{PVT},\ i}$. Then, the generated power by the PV/T system of each peer is found:

$$P_{\mathrm{PVT},i}(t)=A_{\mathrm{PVT},i}\eta_{\mathrm{PVT},i}(t)G_t(t) \tag{9.8}$$

where $G_t$ is the total solar radiation and $\eta_{\mathrm{PVT}}$ is the efficiency of the PV/T panel.

The efficiency of PV/T panels is a function of cell temperature. In this study, Eq. (9.9) is utilized for the calculation of the cell temperature. Then, PV/T efficiency is determined by Eq. (9.10) [23].

$$T_{\mathrm{cell}}(t)=30+0.0175(G_t(t)-300)+1.14(T_{\mathrm{a}}(t)-25) \tag{9.9}$$

$$\eta_{\mathrm{PVT},i}(t)=\eta_{\mathrm{ref}}(1-\beta(T_{\mathrm{cell}}(t)-T_{\mathrm{ref}}) \tag{9.10}$$

$T_{\mathrm{cell}}$, $T_{\mathrm{a}}$ and $T_{\mathrm{ref}}$ are PV cell, ambient and reference temperatures. $\eta_{\mathrm{ref}}$ is also the PV/T reference electrical efficiency.

The PV/T generated heat can also be calculated as follows [23]:

$$Q_{\mathrm{PVT},i}(t)=A_{\mathrm{PVT},i}\left[0.9G_t(t)\left(1-\eta_{\mathrm{PVT},i}(t)\right)-U_{\mathrm{L}}(T_{\mathrm{cell}}(t)-T_{\mathrm{a}}(t))\right] \tag{9.11}$$

where $U_{\mathrm{L}}$ denotes the heat loss coefficient of the PV/T panel.

#### 9.4.1.2 Energy Balance and Sharing Mechanism

The power balance of each peer can be stated as:

$$P_{\mathrm{PVT},i}(t)+P_{\mathrm{ICE}.i}(t)+P^{\mathrm{buy}}_{\mathrm{GRID},i}(t)+P^{\mathrm{buy}}_{\mathrm{P2P},i}(t)-P_{\mathrm{dem},i}(t)-P^{\mathrm{sell}}_{\mathrm{GRID},i}(t)-P^{\mathrm{sell}}_{\mathrm{P2P},i}(t)=0 \tag{9.12}$$

where $P_{\mathrm{PVT}}$ and $P_{\mathrm{ICE}}$ are the generated power by the PV/T and ICE. Also. $P^{\mathrm{buy}}_{\mathrm{GRID}}$ and $P^{\mathrm{buy}}_{\mathrm{P2P}}$ ($P^{\mathrm{sell}}_{\mathrm{GRID}}$ and $P^{\mathrm{sell}}_{\mathrm{P2P}}$) are the power imported from (exported to) the grid and the P2P market, respectively.

A heating and cooling balance should also be stated for each peer, which is defined by Eqs. (9.13) and (9.14), respectively.

$$Q_{\mathrm{PVT},i}(t)+Q_{\mathrm{ICE},i}(t)+Q_{\mathrm{b},i}(t)+Q^{\mathrm{import}}_{\mathrm{P2P-h},i}(t)-Q_{\mathrm{dem},i}(t)-Q^{\mathrm{export}}_{\mathrm{P2P-h},i}(t)-Q_{\mathrm{ex},i}(t)=0 \tag{9.13}$$

$$Q_{\mathrm{AC},i}(t)+Q_{\mathrm{EC},i}(t)+Q^{\mathrm{import}}_{\mathrm{P2P-c},i}(t)-Q_{\mathrm{c},i}(t)-Q^{\mathrm{export}}_{\mathrm{P2P-c},i}(t)=0 \tag{9.14}$$

where $Q_{\mathrm{PVT}}$, $Q_{\mathrm{ICE}}$ and $Q_{\mathrm{b}}$ denote generated heat by PV/T, ICE, and boiler, respectively. $Q_{\mathrm{ex}}$ denotes the excess heat that is exhausted. $Q_{\mathrm{P2P-h}}^{\mathrm{import}}$ ($Q_{\mathrm{P2P-h}}^{\mathrm{export}}$) is the heating imported from (exported to) the P2P system. $Q_{\mathrm{P2P-c}}^{\mathrm{import}}$ ($Q_{\mathrm{P2P-c}}^{\mathrm{export}}$) is also the cooling imported from (exported to) the P2P system.

In the context of the energy sharing mechanism, Eqs. (9.15) to (9.18) are used to ensure that the power/heating/cooling import/export are not carried out simultaneously.

$$\gamma_{\mathrm{GRID},i}^{\mathrm{buy}}(t) + \gamma_{\mathrm{GRID},i}^{\mathrm{sell}}(t) \leq 1 \tag{9.15}$$

$$\gamma_{\mathrm{P2P},i}^{\mathrm{buy}}(t) + \gamma_{\mathrm{P2P},i}^{\mathrm{sell}}(t) \leq 1 \tag{9.16}$$

$$\gamma_{\mathrm{P2P-h},i}^{\mathrm{import}}(t) + \gamma_{\mathrm{P2P-h},i}^{\mathrm{export}}(t) \leq 1 \tag{9.17}$$

$$\gamma_{\mathrm{P2P-c},i}^{\mathrm{import}}(t) + \gamma_{\mathrm{P2P-c},i}^{\mathrm{export}}(t) \leq 1 \tag{9.18}$$

Where $\gamma$ is a binary value expressing the status of energy import/export.

For each peer, it is considered that the priority for cooling supply is for the AC. Hence, when the nominal capacity of the AC is greater than cooling demand, the EC operation and cooling import is not carried out:

$$Q_{\mathrm{AC},i}(t) = Q_{\mathrm{c},i}(t) + Q_{\mathrm{P2P-c},i}^{\mathrm{export}}(t) \tag{9.19}$$

On the other hand, when the AC capacity of each peer does not suffice the cooling demand, the deficiency is met by P2P import and then the EC. In this state, the AC will work on its rated capacity:

$$Q_{\mathrm{AC-nom},i} + Q_{\mathrm{EC},i}(t) + Q_{\mathrm{P2P-c},i}^{\mathrm{import}}(t) = Q_{\mathrm{c},i}(t) \tag{9.20}$$

After a whole year of simulation, the EC capacity for each peer is found as follows:

$$Q_{\mathrm{EC-nom},i} = \max\{Q_{\mathrm{EC},i}(t)\} \tag{9.21}$$

As far as the heating sector is concerned, the generated power by the PV/T and ICE is prior to other sources. Therefore, when it is sufficient:

$$Q_{\mathrm{PVT},i}(t) + Q_{\mathrm{ICE},i}(t) = Q_{\mathrm{dem},i}(t) + Q_{\mathrm{P2P-h},i}^{\mathrm{export}}(t) + Q_{\mathrm{ex},i}(t) \tag{9.22}$$

In case the PV/T and ICE-generated heat does not suffice the heating demand, the deficiency is supplied by P2P import and then the boiler:

$$Q_{\mathrm{PVT},i}(t) + Q_{\mathrm{ICE},i}(t) + Q_{\mathrm{b},i}(t) + Q_{\mathrm{P2P-h},i}^{\mathrm{import}}(t) = Q_{\mathrm{dem},i}(t) \tag{9.23}$$

Here, the boiler capacity of each peer is determined as follows:

$$Q_{\mathrm{b-nom},i} = \max\{Q_{\mathrm{b},i}(t)\} \tag{9.24}$$

The same paradigm is true for the power sharing. When the ICE and PV/T power of peer $i$th is enough to meet the load, then no import is carried out, neither from the P2P network nor the grid:

$$P_{\mathrm{PVT},i}(t) + P_{\mathrm{ICE}.i}(t) = P_{\mathrm{dem},i}(t) + P^{\mathrm{sell}}_{\mathrm{GRID},i}(t) + P^{\mathrm{sell}}_{\mathrm{P2P},i}(t) \tag{9.25}$$

On the other hand, when the PV/T and ICE power does not suffice the power demand of the peer, the deficiency is met first by P2P import and then the grid purchase:

$$P_{\mathrm{PVT},i}(t) + P_{\mathrm{ICE}.i}(t) + P^{\mathrm{buy}}_{\mathrm{GRID},i}(t) + P^{\mathrm{buy}}_{\mathrm{P2P},i}(t) = P_{\mathrm{dem},i}(t) \tag{9.26}$$

#### 9.4.1.3 Objective Function

The objective function is defined as the total annual operating costs ratio as in Eq. (9.27) and is maximized within the optimization:

$$\max \mathrm{AOCR} = \left(1 - \frac{\mathrm{AOC_{P2p}}}{\mathrm{AOC_{ref}}}\right) \tag{9.27}$$

In Eq. (9.27), $\mathrm{AOC_{ref}}$ denotes the annual operating costs of the community, when no multi-energy P2P system is considered, i.e., the power, heating, and cooling of peers are met by the grid, gas boiler, and EC, respectively. Hence, $\mathrm{AOC_{ref}}$ is calculated as follows:

$$\mathrm{AOC_{ref}} = \sum_i \left(C^{\mathrm{op}}_{\mathrm{EC},i} + C^{\mathrm{op}}_{\mathrm{b},i} + C^{\mathrm{buy}}_{\mathrm{grid},i} + C^{\mathrm{buy}}_{\mathrm{NG},i}\right) \tag{9.28}$$

In Eq. (9.27), $\mathrm{AOC_{P2p}}$ is the annual operating costs of the community, regarding the multi-energy P2P system. Hence, it is determined as follows:

$$\mathrm{AOC_{P2p}} = \sum_i \begin{pmatrix} C^{\mathrm{op}}_{\mathrm{ICE},i} + C^{\mathrm{op}}_{\mathrm{PVT},i} + C^{\mathrm{op}}_{\mathrm{b},i} + C^{\mathrm{op}}_{\mathrm{EC},i} + C^{\mathrm{op}}_{\mathrm{AC},i} + C^{\mathrm{buy}}_{\mathrm{grid}i} + C^{\mathrm{buy}}_{\mathrm{P2P},i} \\ + C^{\mathrm{buy}}_{\mathrm{NG},i} - C^{\mathrm{sell}}_{\mathrm{grid}i} - C^{\mathrm{sell}}_{\mathrm{P2P},i} \end{pmatrix} \tag{9.29}$$

### 9.4.2 *Optimization Results*

The model proposed in Sect. 9.4.1 was optimized using genetic algorithm. The specifications of the algorithm used are presented in Table 9.3. The optimal values for the multi-energy system components of each peer are presented in Table 9.4. As can be seen in Table 9.4, the optimal capacity of ICE in Peer 1 is equal to 345 kW, while this value for Peer 2 is equal to 100 kW. Also, the number of PVT panels in peer 1 is equal to the maximum possible value, i.e., 250, while in peer 2, it is equal to the minimum possible value, i.e., zero. A large share of the cooling load in Peer 1 is provided by the absorption chiller, and on the other hand, in Peer 2, most of the required cooling is met by the electric chiller.

Local power generating devices in each peer include PVT panels and ICE. PVT output is a function of environmental factors such as radiation and ambient temperature, while ICE output is determined based on peer power demand and local market. Figure 9.6 shows the daily profile of local power generation in two buildings for the two middle days of summer and winter.

**Table 9.3** Genetic algorithm characteristics

| | |
|---|---|
| Population size | 50 |
| Number of generations | 100 |
| Crossover percentage (%) | 80 |
| Mutation percentage (%) | 30 |
| Selection method | Roulette wheel |
| Selection pressure | 8 |

**Table 9.4** Optimal component sizes of peer 1 and 2

| Optimal component sizes | ICE (kW) | PVT (#) | Boiler (kW) | AC (kW) | EC (kW) |
|---|---|---|---|---|---|
| Peer 1 (Commercial building) | 345 | 250 | 578 | 714 | 175 |
| Peer 2 (Office building) | 100 | 0 | 762 | 100 | 1152 |

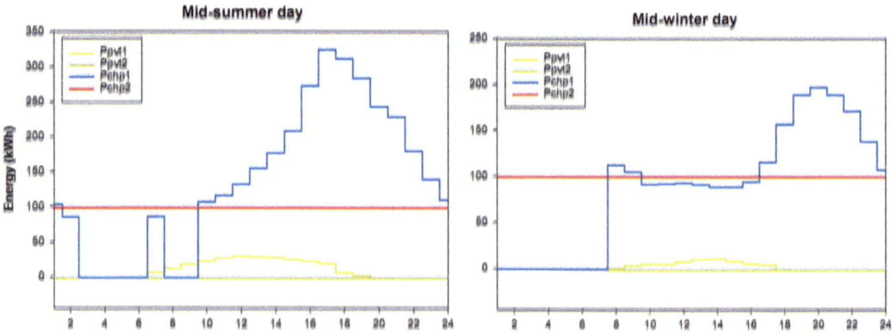


**Fig. 9.6** PVT and ICE power generation profile of peers

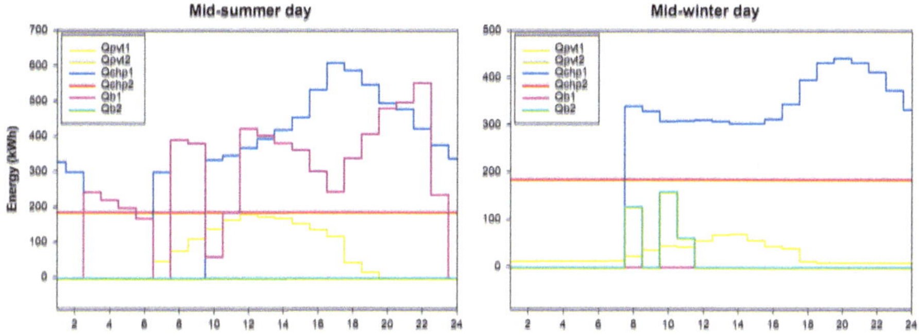


**Fig. 9.7** PVT, ICE, and boiler heat generation profile of peers

The production power of the PVT system in peer 1 is clearly lower on mid-winter day than on mid-summer day. Due to the absence of PVT system in Peer 2, the corresponding power is zero. The output of ICE in Peer 2 is at full capacity on both mid-winter and mid-summer days, while the output power of ICE in Peer 1 is clearly a function of demand.

The production of heat required by each peer as well as the integrated heat network is provided by PVT, ICE recovered heat and boiler. Figure 9.7 depicts the heat produced by these components in two peers and during the mid-winter and mid-summer days. As it can be seen, the heat produced by ICE in Peer 2 is constant due to constant operation at maximum capacity. On the other hand, the heat of ICE in peer 1 is a function of its performance status. In the mid-summer day, when the ICE is not working, the boiler 1 is turned on, while the boiler 2 is completely off. On the other hand, in the mid-winter day, boiler 2 operates in some hours and boiler 1 is off.

The required power of each peer is supplied either through local production (PVT or ICE) or through import from another peer or the power grid. Figure 9.8 shows the daily power profile in two peers along with the power supply source (Column a: Peer 1; Column b: Peer 2).

In Peer 1 and during the mid-summer day, there is no P2P import and most of the required power is provided through local production. In the early hours of the day and when the ICE is off, power is supplied through the grid. Although the priority of supply is through P2P trade, because the production power of peer 2 is lower than its internal demand, the power deficit in peer 1 is completely supplied from the grid. This is also true in winter, where in the early hours of the day, the main demand of Peer 1 is met by the network.

In Peer 2, the power requirement is provided by all three methods of grid, local production, and P2P trading. Whether in mid-summer or mid-winter day, the production capacity of ICE is constant (rated capacity); and in the middle of the day, part of the power is provided through the P2P network. Rest of the load is also imported from the grid.

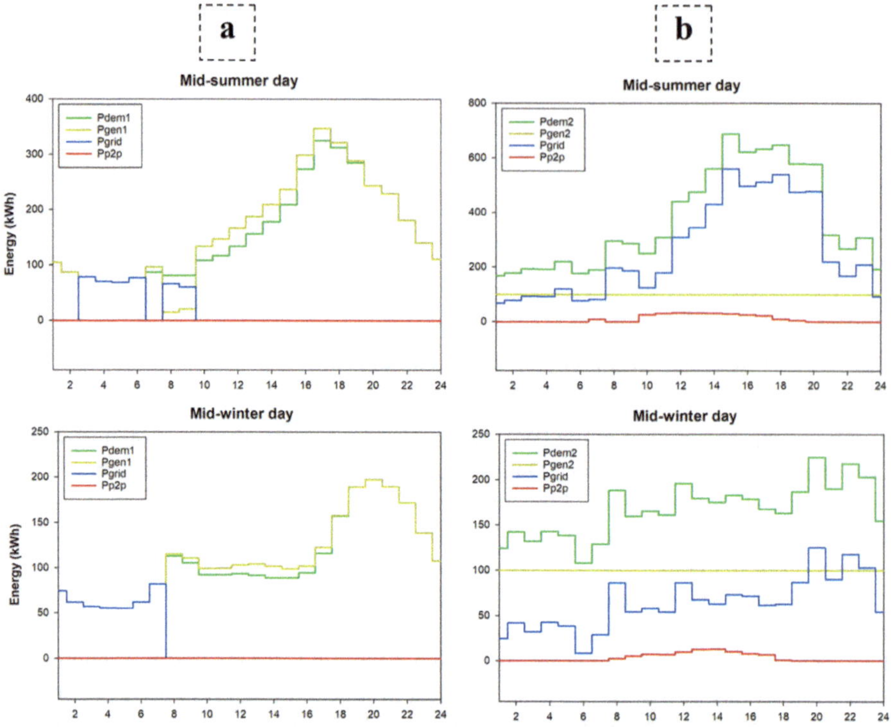


**Fig. 9.8** Daily power profile in two peers along with the power supply source (Column **a**: Peer 1; Column **b**: Peer 2)

Like the power sector, part of the heat required by the peers is supplied through local production (ICE, PVT, or boiler) and its deficiency is imported through the P2P heat exchange network. Figure 9.9 shows the daily profile of heat demand of peers along with sources of heat supply (Column a: Peer 1; Column b: Peer 2).

In peer 1 and mid-summer day, there is a high heat demand due to the absorption chiller. A major part of this heat is supplied through local production and boilers, and the deficiency is imported from peer 2. In the winter, the local production of heat (mainly heat recovered from ICE) is more than the domestic demand, so the boiler does not work; During these hours, heat exhaust is high. In the early hours of the day, a major part of the demand is provided through the P2P network.

In peer 2 and on mid-summer day, ICE's heat production exceeded the internal demand, and as a result, its surplus is injected into the P2P network. In some hours of the day, heat loss has also occurred. In mid-winter day and during office hours, most of the heating needs are supplied through ICE and the P2P network, and the boiler also operates in some hours. As it is clear, in the early hours of the day, the heat exchange was from peer 2 to 1 and then vice versa, i.e., from peer 1 to 2. This clearly confirms the optimal operation and benefit of a multi-energy P2P system.

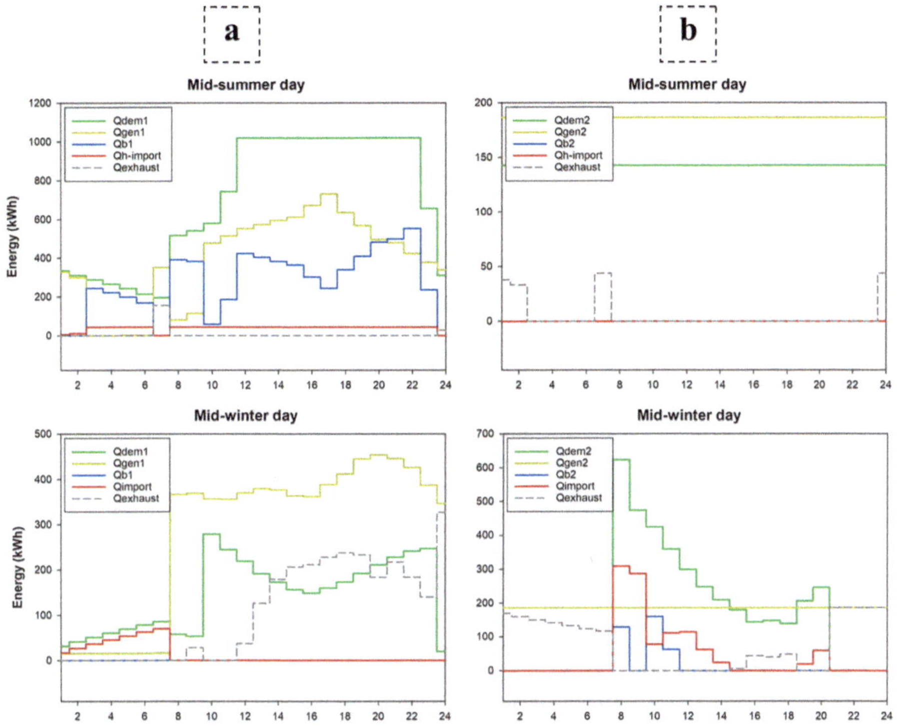


**Fig. 9.9** Daily profile of heat demand of peers along with sources of heat supply (Column **a**: Peer 1; Column **b**: Peer 2)

## 9.5 Conclusion

Energy systems are increasingly transforming and the structure of their components and actors are changing. Within the past two decades, energy systems have been changing from the conventional unilateral and vertical structure of the past to the bilateral and horizontal structure. The use of demand-side management methods, the expansion of distributed energy generation sources, the development of renewable energy resources, the improvement of energy storage technologies, and the emergence of new energy consumers such as electric vehicles, have all accelerated the process of this restructuring.

As one of the latest transition trends in energy systems, P2P energy trading is considered an efficient solution in order to improve the efficiency of the energy system and increase the participation of prosumers/consumers in regulating and managing the market. A prominent approach in the P2P energy trading market is the integration of multi-energy systems in these networks. A multi-energy P2P system, while enjoying the benefits of a P2P market, can improve the efficiency of the energy system by integrating various energy carriers, various energy conversion systems, and various storage components.

## References


1. Muhsen, H., et al. (2022). Business model of peer-to-peer energy trading: A review of literature. *Sustainability, 14*(3), 1616.
2. Yousefi, H., & Ghodusinejad, M. H. (2017). Feasibility study of a hybrid energy system for emergency off-grid operating conditions. *Majlesi Journal of Electrical Engineering, 11*(3).
3. Sousa, T., et al. (2019). Peer-to-peer and community-based markets: A comprehensive review. *Renewable and Sustainable Energy Reviews, 104*, 367–378.
4. Zhou, Y., et al. (2020). State-of-the-art analysis and perspectives for peer-to-peer energy trading. *Engineering, 6*(7), 739–753.
5. Mohammadi, M., et al. (2017). Energy hub: From a model to a concept–A review. *Renewable and Sustainable Energy Reviews, 80*, 1512–1527.
6. Wang, N., et al. (2022). A peer-to-peer market mechanism incorporating multi-energy coupling and cooperative behaviors. *Applied Energy, 311*, 118572.
7. Zhong, X., et al. (2022). Optimal energy management for multi-energy multi-microgrid networks considering carbon emission limitations. *Energy, 246*, 123428.
8. Jalilian, F., et al. (2022). Multi-energy microgrids: An optimal despatch model for water-energy nexus. *Sustainable Cities and Society, 77*, 103573.
9. Javadi, M. S., et al. (2022). Transactive energy framework in multi-carrier energy hubs: A fully decentralized model. *Energy, 238*, 121717.
10. Zhang, H., et al. (2022). Parametric optimization-based peer-to-peer energy trading among commercial buildings considering multiple energy conversion. *Applied Energy, 306*, 118040.
11. Daneshvar, M., Mohammadi-Ivatloo, B., & Zare, K. (2022). A novel transactive energy trading model for modernizing energy hubs in the coupled heat and electricity network. *Journal of Cleaner Production, 344*, 131024.
12. Zhang, G., et al. (2022). A multi-agent deep reinforcement learning approach enabled distributed energy management schedule for the coordinate control of multi-energy hub with gas, electricity, and freshwater. *Energy Conversion and Management, 255*, 115340.
13. Daneshvar, M., et al. (2020). Chance-constrained models for transactive energy management of interconnected microgrid clusters. *Journal of Cleaner Production, 271*, 122177.
14. Davoudi, M., & Moeini-Aghtaie, M. (2022). Local energy markets design for integrated distribution energy systems based on the concept of transactive peer-to-peer market. *IET Generation, Transmission & Distribution, 16*(1), 41–56.
15. Liu, J., Yang, H., & Zhou, Y. (2021). Peer-to-peer trading optimizations on net-zero energy communities with energy storage of hydrogen and battery vehicles. *Applied Energy, 302*, 117578.
16. Davoudi, M., Moeini-Aghtaie, M., & Ghorani, R. (2021). Developing a new framework for transactive peer-to-peer thermal energy market. *IET Generation, Transmission & Distribution, 15*(13), 1984–1995.
17. Khorasany, M., et al. (2021). Transactive energy framework for optimal energy management of multi-carrier energy hubs under local electrical, thermal, and cooling market constraints. *International Journal of Electrical Power & Energy Systems, 129*, 106803.
18. Li, L., & Zhang, S. (2021). Peer-to-peer multi-energy sharing for home microgrids: An integration of data-driven and model-driven approaches. *International Journal of Electrical Power & Energy Systems, 133*, 107243.
19. Nguyen, D. H., & Ishihara, T. (2021). Distributed peer-to-peer energy trading for residential fuel cell combined heat and power systems. *International Journal of Electrical Power & Energy Systems, 125*, 106533.

20. Daneshvar, M., et al. (2020). A novel operational model for interconnected microgrids participation in transactive energy market: A hybrid IGDT/stochastic approach. *IEEE Transactions on Industrial Informatics, 17*(6), 4025–4035.
21. Liu, N., et al. (2018). Distributed energy sharing for PVT-HP prosumers in community energy internet: A consensus approach. *Energies, 11*(7), 1891.
22. Yousefi, H., Ghodusinejad, M. H., & Noorollahi, Y. (2017). GA/AHP-based optimal design of a hybrid CCHP system considering economy, energy and emission. *Energy and Buildings, 138*, 309–317.
23. Yousefi, H., Ghodusinejad, M. H., & Kasaeian, A. (2017). Multi-objective optimal component sizing of a hybrid ICE+ PV/T driven CCHP microgrid. *Applied Thermal Engineering, 122*, 126–138.

# Chapter 10
# Optimal Scheduling of Local Peer-to-Peer Energy Trading Considering Hydrogen Storage System

**Ali Aminlou, Ramin Nourollahi, and Kazem Zare**

## 10.1 Introduction

In recent years, the traditional energy market structure undergoes significant changes. Several factors have contributed to the current change in the energy sector. These factors include the spread of decentralized green energy sources, the rise in prosumers connected to the local electrical grid, and the advancement of communications technology [1]. Consumers who act as producers of renewable energy, such as rooftop solar panels, are connected to the grid and exchange their extra energy. In order to accommodate the new players called peers through various trading methods and balance supply and demand inside the microgrid, the local energy market is experiencing several modifications [2]. By introducing new energy transaction infrastructure, such as the peer-to-peer (P2P) market, the power system is undergoing a transition from its old hierarchical structure to a more decentralized form. A good deal of research and pilot projects have shown how P2P sharing has been proven to be beneficial for prosumers and the grid [3, 4]. P2P energy transactions provide several benefits, including financial gain from market participation, increased system resilience, and decreased carbon emissions [5]. In the P2P energy trading market, every prosumer participating in the market, which is called peers, shares their extra energy generation with the other peers. In the traditional structure of the P2P energy trading, peers shared their resources under a client-server paradigm with the centralized server. However, in recent studies, prosumers share their energy resource with no third party. These approaches can combine to make a hybrid system. One of the well-known P2P system applications is Bitcoin, which is a P2P-based digital currency [6].

---

A. Aminlou (✉) · R. Nourollahi · K. Zare
Faculty of Electrical and Computer Engineering, University of Tabriz, Tabriz, Iran
e-mail: aliaminlou@tabrizu.ac.ir



V. Vahidinasab, B. Mohammadi-Ivatloo (eds.), *Demand-Side Peer-to-Peer Energy Trading*, Green Energy and Technology, https://doi.org/10.1007/978-3-031-35233-1_10

Peer-to-peer energy trading is a new concept in the energy market. In this market structure, peers are individual prosumers that produce their own energy and share the extra energy in the market. P2P energy market requires additional research to improve their market models, information technologies, and operational systems because they are not yet developed systems [7]. The effects of the P2P economy may be examined from several angles. There will be an important impact on social and cultural norms of prosumers around the supply and demand market. The administration and maintenance of P2P networks will cause local training and employment possibilities. Furthermore, a higher degree of social trust will be developed, enhancing energy trading transparency and lowering tricky transactions. The participants will ultimately feel closer to other customers and more attached to the community. The peers must communicate with each other and cooperate to maximize the profit [2, 8]. P2P energy trading concept is a practical scheme that is implemented all over the world. Some of the brilliant examples for this concept are Sonnen Community in Germany, Transactive Grid and Yeloha in the US, Piclo in the UK, and Vandebron in the Netherlands [7, 9–12]. The number of power providers and utilities operating in the German energy market by 2020 was over 2000. Customers want to consume less from the upper grid and be self-sufficient [13]. By using the P2P market in feed-in-tariff policy, the prosumers that are equipped with the PV systems may significantly reduce their costs compared to trading directly with the utility grid [14].

### *10.1.1 Related Works*

In order to motivate individual users, the [15] propose a cooperative game theory framework. Depending on factors like maximum energy consumption and generation and price mechanism, this trading algorithm provides a different priority at each time period. This paper tries to maximize the coalition's social welfare in a Nash equilibrium and makes sure that all the participants benefit from the coalition's policies. This study indicates how trading under a P2P market is an appealing concept that reduces power prices for consumers, boosts the use of renewable energy, and provides prosumers money. It enhances the use of distributed energy resources (DERs) and lowers transmission and distribution losses. In order to increase energy efficiency and save the cost of each peer [16], presents a multi-energy coupling framework. All peers develop their trading strategies depending on the time-of-use tariff pricing system to maximize their earnings. This paper employed a parametric optimization-based to model the trading problem. In this scheme, every peer must consider the other customer's willingness to plan his own strategy. The proposed P2P transaction mechanism can lower buildings' energy costs. This paper shows that the energy trading between various buildings reduces the amount of daily electricity bills by 8.9%. In the following [17] present a P2P token market using the blockchain technology in a smart grid infrastructure. This paper used a fully decentralized approach to model the small-scale peer's transaction. This paper divides the customers into two groups of buyers and sellers. In this paper, peers can negotiate with each other on the energy price and finally engage in smart contract to trade energy with each other. This paper utilizes the primal-dual sub-gradient method to clear the market. The paper [18] employed aggregative game to formulate the P2P market mechanism. This paper guarantees the problem convergence and shows the scalability or work in IEEE 37-bus test-case. This paper also shows that participants and network operator benefit from the P2P market existence. The paper [19] prepares a framework to convert the energy to gas and reduce the Carbon dioxide ($CO_2$) emissions. At the first level, this paper minimizes the $CO_2$ emission and cost. In the second level, $CO_2$ is stored in the Carbon Capture and Storage (CCS) and finally Methane ($CH_4$) is produced by a combination of $CO_2$ and Hydrogen, which is provided by the water electrolyzer. The paper [20] considers the three types of buildings equipped with solar energy generation and hydrogen storage unit. These units are connected to each other and the upper grid. This paper considers the uncertainty of the load and PV generation and shows that the units get revenue by participating in this market. In the following, the paper [21] proposes another P2P market structure in a net-zero energy community. This paper considers the high penetration of residential buildings and renewable generation like PV and wind. Also, this community is equipped with hydrogen vehicle storage. This study shows that this market can reduce the overall cost and imported energy from the grid and decrease the $CO_2$ emission. Hydrogen-fueled electric cars have advanced quickly in recent years. In order to examine the effects, the paper [22] develops the operational model of the coupled power-traffic system. Then, a cooperative architecture of P2P energy trading is developed to make a connection between power network and Hydrogen-fueled electric cars. This paper prevents the arbitrage and employed a dynamic pricing to reduce cost and guarantee transaction fairness.

## 10.2 The Proposed Model

In the traditional structure of a power system, every customer is connected to the main grid and this unit can trade energy with the upper grid. In the new market structure, customers are equipped with the new technologies like renewable energy resources and hydrogen energy storage system. The P2P energy market enables the peers to participate in this market and share energy with each other. This new market structure does not need the third party to manage the transaction between the peers and every customer can individually participate in market and share their resources with other peers.

### *10.2.1 Hydrogen Storage Unit*

Three different parts make up the hydrogen energy storage unit. The water electrolyzer component, hydrogen storage container, and fuel-cell supplies are all included in a hydrogen storage system. The charging process activates by transforming the water to the hydrogen in the electrolyzer. For the green creation of carbon-free hydrogen from electric energy, electrolysis is a potential alternative. The process of separating water into hydrogen and oxygen is known as electrolysis. This reaction occurs in a device known as an electrolyzer. The hydrogen produced in this charging method is stored in a hydrogen storage container. The reverse process for discharging energy is activated by a fuel-cell. In the fuel-cell, hydrogen is converted to electricity and produces energy [23].

### 10.2.2 Peer-to-Peer Energy Market

As shown in Fig. 10.1, the proposed model consists of three individual buildings. The first and second buildings are equipped with solar energy generation and residential load. The third peer is equipped with a hydrogen energy storage unit. Energy storage units let the peers overcome the disadvantages of the PV system uncertainty. When the PV generation is high, the energy storage unit charges the hydrogen tank and stores the extra energy. The storage unit can supply the demand when the PV generation is not available. Every unit in this model acts as a single player and the load profile of each unit differs from the others. The P2P energy

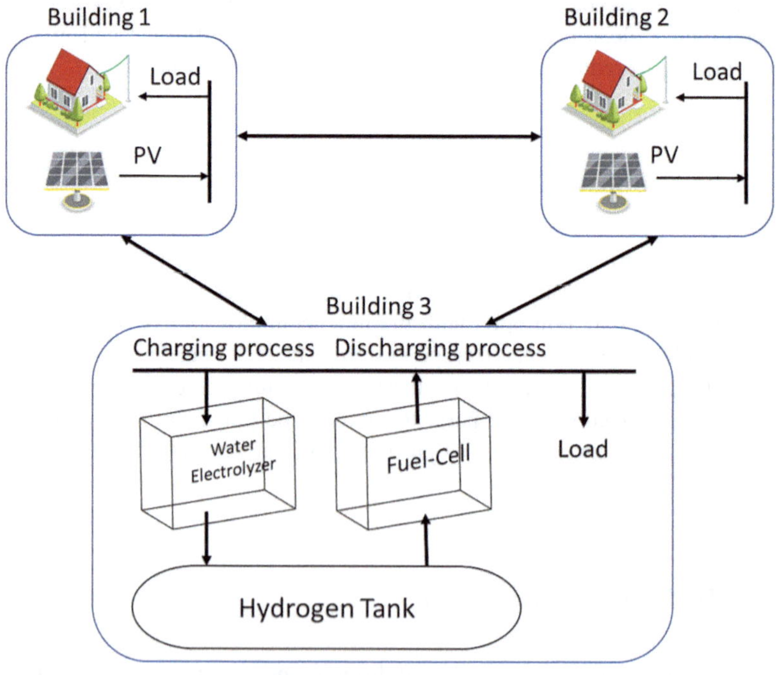


**Fig. 10.1** Schematic overview of a P2P market in an energy community

market lets the peers participating in the market benefit from each other. For example, the first and second buildings can benefit from the third building by sharing the hydrogen storage system. The P2P market also reduce the investment cost because the building does not need to install the new equipment and they share their facilities and benefit from each other.

## 10.3 Problem Formulation

This paper considers the COST as an overall cost in a year. The first term of the overall cost function shown in (10.1) is pointing to the investment cost. In this equation TInvC shows the total investment cost and AInvC pointing the annual investment cost. In the following, second and third term show the maintenance cost and salvage value. In this relation, SV and MaintC shows the salvage value and maintenance cost. The yearly investment cost is defined in the (10.2). The investment cost is dependent on the discount rate (Disr) and the system life span (lifes). The yearly salvage values (YSV) which is shown in (10.3) demonstrate the book value of the system after calculation of all depreciation.

$$\mathrm{COST} = \mathrm{TInvC} \times \mathrm{AInvC} + \mathrm{MaintC} \times 365 - \mathrm{SV} \times \mathrm{YSV} \tag{10.1}$$

$$\mathrm{AInvC} = \frac{\left(\mathrm{Disr} \times (1 + \mathrm{Disr})^{\mathrm{lifes}}\right)}{(1 + \mathrm{Disr})^{\mathrm{lifes}} - 1} \tag{10.2}$$

$$\mathrm{YSV} = \frac{\mathrm{Disr}}{(1 + \mathrm{Disr})^{\mathrm{lifes}} - 1} \tag{10.3}$$

In the following, annual cost of each energy resource is calculated as follows. The relation (10.4) and (10.5) displays the annual cost of the PV generation and the transmission network cost between the peers. In these equations, $P_{\mathrm{pv1}}$, $P_{\mathrm{pv2}}$ shows the power produced by the PV in the first and second customer and $\lambda_{\mathrm{pv}}$ shows the price for solar panels ($/kW). Salvage value of solar panels is shown by $\mathrm{SV}_{\mathrm{pv}}$ and daily maintenance cost is shown by $\mathrm{MC}_{\mathrm{pv}}$ ($/kW).

$$\mathrm{Cost}_{\mathrm{pv}} = \left(P_{\mathrm{pv1}} + P_{\mathrm{pv2}}\right) \times \left(\lambda_{\mathrm{pv}} \times \mathrm{AInvC} - \mathrm{SV}_{\mathrm{pv}} \times \mathrm{YSV} + \mathrm{MC}_{\mathrm{pv}} \times 365\right) \tag{10.4}$$

$$\mathrm{Cost}_{\mathrm{pl}} = \left(\mathrm{Pl}_n^{12} + \mathrm{Pl}_n^{23} + \mathrm{Pl}_n^{31}\right) \times \left(\lambda_{\mathrm{pl}} \times \mathrm{AInvC} - \mathrm{Sv}_{\mathrm{pl}} \times \mathrm{YSV} + \mathrm{Mc}_{\mathrm{pl}} \times 365\right) \tag{10.5}$$

The cost of transmission lines between the peers is shown in (10.5). In this equation $\mathrm{Pl}_n$ shows the nominal power of lines. $\lambda_{\mathrm{pl}}$ demonstrate the price of power lines ($/kW).

In the following, the Annualized cost of fuel-cell, electrolyzer, and hydrogen tank are checked out.

$$\mathrm{Cost}_{\mathrm{fc}} = (P_{\mathrm{fc}}) \times (\lambda_{\mathrm{fc}} \times \mathrm{AInvC} - \mathrm{Sv}_{\mathrm{fc}} \times \mathrm{YSV} + \mathrm{Mc}_{\mathrm{fc}} \times 365) \tag{10.6}$$

$$\mathrm{Cost}_{\mathrm{el}} = (P_{\mathrm{el}}) \times (\lambda_{\mathrm{el}} \times \mathrm{AInvC} - \mathrm{Sv}_{\mathrm{el}} \times \mathrm{YSV} + \mathrm{Mc}_{\mathrm{el}} \times 365) \tag{10.7}$$

$$\mathrm{Cost}_{\mathrm{hyt}} = \left(P_{\mathrm{hyt}}\right) \times \left(\lambda_{\mathrm{hyt}} \times \mathrm{AInvC} - \mathrm{Sv}_{\mathrm{hyt}} \times \mathrm{YSV} + \mathrm{Mc}_{\mathrm{hyt}} \times 365\right) \tag{10.8}$$

$$\mathrm{Tcost} = e = \mathrm{Cos}_{\mathrm{PV}} + \mathrm{Cos}_{\mathrm{PL}} + \mathrm{Cos}_{\mathrm{fc}} + \mathrm{Cos}_{\mathrm{el}} + \mathrm{Cos}_{\mathrm{hyt}} \tag{10.9}$$

The Eq. (10.6) evaluates the fuel-cell costs. Also $P_{\mathrm{fc}}$, $\lambda_{\mathrm{fc}}$ shows the fuel-cell power (kW) and price of fuel-cell ($/kW) respectively. Annual cost of the water electrolyzer and hydrogen tank is shown in (10.7) and (10.8). In equations, $P_{\mathrm{el}}$ and $P_{\mathrm{hyt}}$ shows the power of water electrolyzer (kW) and Capacity of hydrogen tank (kg). Also $\lambda_{\mathrm{el}}$ and $\lambda_{\mathrm{hyt}}$ shows the water electrolyzer and hydrogen tank price. The objective function of this problem is minimizing the total cost which shown in (10.9).

As shown in Fig. 10.1, the first and second peer are equipped with PV energy production system. This building can trade energy with other peers. The (10.10) and (10.12) define power balance for them. The solar energy generation is limited by (10.11) and (10.13) to the planned power of the system.

$$P^{b1}_{\gamma,k,h} + P^{12}_{\gamma,k,h} + P^{13}_{\gamma,k,h} - P^{s1}_{\gamma,k,h} = 0 \quad \forall\gamma, \in S, k \in Y, h \in T \tag{10.10}$$

$$P^{s1}_{\gamma,k,h} \leq P_{\mathrm{pv1}} \quad \forall\gamma, \in S, k \in Y, h \in T \tag{10.11}$$

$$P^{b2}_{\gamma,k,h} + P^{23}_{\gamma,k,h} + P^{21}_{\gamma,k,h} - P^{s2}_{\gamma,k,h} = 0 \quad \forall\gamma, \in S, k \in Y, h \in T \tag{10.12}$$

$$P^{s2}_{\gamma,k,h} \leq P_{\mathrm{pv1}} \quad \forall\gamma, \in S, k \in Y, h \in T \tag{10.13}$$

In this chapter, the set of scenarios, seasons, and time periods is defined by $\gamma$, $k$ and $h$ respectively. $P^{b1}_{\gamma,k,h}$ and $P^{b2}_{\gamma,k,h}$ shows the load of the building 1 and 2. $P^{12}_{\gamma,k,h}$ demonstrate the power that is transferred from the peer 1 to the peer 2. $P^{s1}_{\gamma,k,h}$ and $P^{s2}_{\gamma,k,h}$ shows the PV generation in time $t$ and season $y$ under the scenario $s$.

As described above, the last peer is equipped with the hydrogen energy storage. The power balance of this peer is defined by (10.14). In this equation $P^{b3}_{\gamma,k,h}$ shows the load of the building 3. And $P^{\mathrm{el}}_{\gamma,k,h}$ and $P^{\mathrm{fc}}_{\gamma,k,h}$ define the stored energy by electrolyzer and discharged by the fuel-cell process.

$$P^{b3}_{\gamma,k,h} + P^{31}_{\gamma,k,h} + P^{32}_{\gamma,k,h} + P^{\mathrm{el}}_{\gamma,k,h} - P^{\mathrm{fc}}_{\gamma,k,h} = 0 \quad \forall\gamma, \in S, k \in Y, h \in T \tag{10.14}$$

$$P^{\mathrm{el}}_{\gamma,k,h} \leq P_{\mathrm{el}} \quad \forall\gamma, \in S, k \in Y, h \in T \tag{10.15}$$

$$P^{\mathrm{fc}}_{\gamma,k,h} \leq P_{\mathrm{fc}} \quad \forall\gamma, \in S, k \in Y, h \in T \tag{10.16}$$

The transactive energy between the peers is defined through (10.17)–(10.19). This equation shows that the transferred energy from every peer to other peers should be equal to the received energy in the destination.

$$P^{12}_{\gamma,k,h} + P^{21}_{\gamma,k,h} = 0 \quad \forall\gamma, \in S, k \in Y, h \in T \tag{10.17}$$

$$P^{23}_{\gamma,k,h} + P^{32}_{\gamma,k,h} = 0 \quad \forall\gamma, \in S, k \in Y, h \in T \tag{10.18}$$

$$P^{31}_{\gamma,k,h} + P^{13}_{\gamma,k,h} = 0 \quad \forall\gamma, \in S, k \in Y, h \in T \tag{10.19}$$

In the following, the energy transmission lines nominal capacity is defined by (10.20)–(10.22).

$$\left|P^{12}_{\gamma,k,h}\right| \leq \text{Pl}^{12}_{n} \quad \forall\gamma, \in S, k \in Y, h \in T \tag{10.20}$$

$$\left|P^{13}_{\gamma,k,h}\right| \leq \text{Pl}^{31}_{n} \quad \forall\gamma, \in S, k \in Y, h \in T \tag{10.21}$$

$$\left|P^{23}_{\gamma,k,h}\right| \leq \text{Pl}^{23}_{n} \quad \forall\gamma, \in S, k \in Y, h \in T \tag{10.22}$$

In the following equations, the hydrogen tank formulation is checked out. The hydrogen storage system can generate energy by using the hydrogen in fuel-cell and consume energy by water electrolyzer system. Doing both of this process in same time is not logical. So, the following Eq. (10.23) prevents the charging/discharging in the same time slot. In this equation, $C^{h}_{\gamma,k,h}$ and $G^{h}_{\gamma,k,h}$ show the hydrogen consumption by fuel-cell and hydrogen generated by the water electrolyzer.

$$\begin{cases} \text{if} \rightarrow C^{h}_{\gamma,k,h} > 0 \Rightarrow G^{h}_{\gamma,k,h} = 0 \\ \text{if} \rightarrow G^{h}_{\gamma,k,h} > 0 \Rightarrow C^{h}_{\gamma,k,h} = 0 \\ \forall\gamma, \in S, k \in Y, h \in T \end{cases} \tag{10.23}$$

The hydrogen consumption and generation process modeling are shown in (10.24) and (10.25). In these equations, $\text{cf}^{e}_{h}$ is the conversion factor between hydrogen and electricity.

$$C^{h}_{\gamma,k,h} = P^{\text{fc}}_{\gamma,k,h} \times \text{cf}^{e}_{h} \quad \forall\gamma, \in S, k \in Y, h \in T \tag{10.24}$$

$$G^{h}_{\gamma,k,h} = P^{\text{we}}_{\gamma,k,h} \times \text{cf}^{e}_{h} \quad \forall\gamma, \in S, k \in Y, h \in T \tag{10.25}$$

In (10.26) the energy that is stored in the hydrogen tank is calculated in every time slot. $H^{s}_{\gamma,k,h}$ shows the stored hydrogen (kg).

$$H^{s}_{\gamma,k,h} = H^{s}_{\gamma,k,h} + G^{h}_{\gamma,k,h} - C^{h}_{\gamma,k,h} \quad \forall\gamma, \in S, k \in Y, h \in T \tag{10.26}$$

$$\eta_{s,y}^{hs} = \frac{\sum_{t \in T} C_{\gamma,k,h}^{h}}{\sum_{t \in T} G_{\gamma,k,h}^{h}} \quad \forall \gamma, \in S, k \in Y, h \in T \tag{10.27}$$

$$H_{\gamma,k,h}^{s} \leq Pn_{ht} \quad \forall \gamma, \in S, k \in Y, h \in T \tag{10.28}$$

Finally, (10.27) calculates the energy storage efficiency and (10.28) demonstrates that stored hydrogen is limited by hydrogen tank capacity.

### 10.3.1 The Risk Management Methodology

To protect the system against situations that are not desirable, the risks of uncertainty should be taken into consideration while solving the optimum scheduling problem. The conditional value-at-risk (CVaR) approach, which derives from the VaR methodology, is used to quantify the stochastic programming risk. The CVaR is known as mean value-at-risk or average excess loss. The CVaR brings advantages like computational advantages when a very large number of scenarios are considered in optimization problem [24]. By computing the mean of the loss exceeding the value of VaR, CVaR creates a more accurate signal of future losses that might exceed the assumed confidence level. The following calculation presents the formulation of CVaR.

$$\text{Min}(1-\beta)\text{Tcost} - \beta\left(\zeta + \frac{1}{1-\alpha}\sum_{s}\Pi_s \times k_s\right) \tag{10.29}$$

$$\text{Subject to :}$$

$$\text{Tcost}_s - \zeta \leq k_s \tag{10.30}$$

$$k_s \geq 0 \tag{10.31}$$

$$(10.10)\text{–}(10.28) \tag{10.32}$$

In this formulation $\text{Tcost}_s$ shows the total cost in each scenarios and $\beta$ and $\alpha$ demonstrate the coefficient of risk and confidence level, respectively. This chapter consider the confidence level to 95%. In the Eq. (10.30), if the results of the $\text{Tcost}_s - \xi$ become negative, then the $k_s$ get the zero value, otherwise $k_s$ will be equal to the $\text{Tcost}_s - \xi$.

## 10.4 Case Study and Results

The residential area is assessed under a variety of precarious circumstances for the purpose of this case study. In order to protect themselves from the unpredictability of the electrical market, there are three homes that have been fitted with alternative forms of power generation, such as solar panels and fuel storage. On the other hand, users in residential areas are often required to be connected to the primary power system since the available energy supplies may not supply the demand for electricity in the residential grid. The price of electricity might be much more expensive than usual during peak hours. As a consequence of this, we are going to have to take the required measures in order to deal with the high cost of energy under the present conditions. Every single house in the area under consideration has both a fuel cell system and a renewable energy source installed in their homes. In order to expedite the transfer of energy from one peer to another, particularly during peak hours, these buildings are connected with one another using P2P transactions. This strategy may be helpful in assisting residential areas lower the amount of energy they need from the electrical grid. The paper [20] provides information that is comprehensive and detailed on the system that is being considered. The proposed system is examined in the context of unpredictably occurring circumstances, such as rising utility prices and interruptions in power supply. As a result, the investigation is formulated by employing the conditional value-at-risk methodology, which is a method of doing risk analysis. The recommended model was examined in this way in both the No-risk and the With-risk modes, and the conclusions that resulted from doing so are presented further down in this section.

Figures 10.2, 10.3, 10.4, and 10.5 illustrate the energy that is transferred along these lines during transmission. This performance in between the lines has been evaluated throughout the course of 1 day and four distinct seasons. Figures 10.2, 10.3, 10.4, and 10.5 illustrate, in order, the outcomes of the first, second, third, and fourth seasons, respectively. Specifically, as can be seen in Fig. 10.2, there has not been a transfer of power between hours 9 and 17 in any of the transactions that occurred between 2 and 3. This is in contrast to the transactions that occurred between 3 and 1. On the other hand, it is crystal clear that the importation of electricity achieves its zenith on lines 3 to 1 at hour 13 when it reaches its maximum point. In general, the amount of energy imported in the model with risk is more than the amount imported in the model without risk, and the bulk of the power trading that took place between the buildings happened between the hours of 9 and 17.

The operation of the components of the system, including the solar systems, fuel cell, hydrogen storage tank, and electrolyzer, is shown in Fig. 10.6. This functionality is demonstrated for both the No-risk model and the With-risk model. To begin, the functioning of the components that make up the hydrogen tank and fuel cell system is exactly the same in both the With-risk model and the No-risk model. In addition, the performance of the electrolyzer system was slightly improved in the No-risk model with 105 kw compared to the performance in the With-risk model. Between the No-risk model and the With-risk model of the first solar system, there is

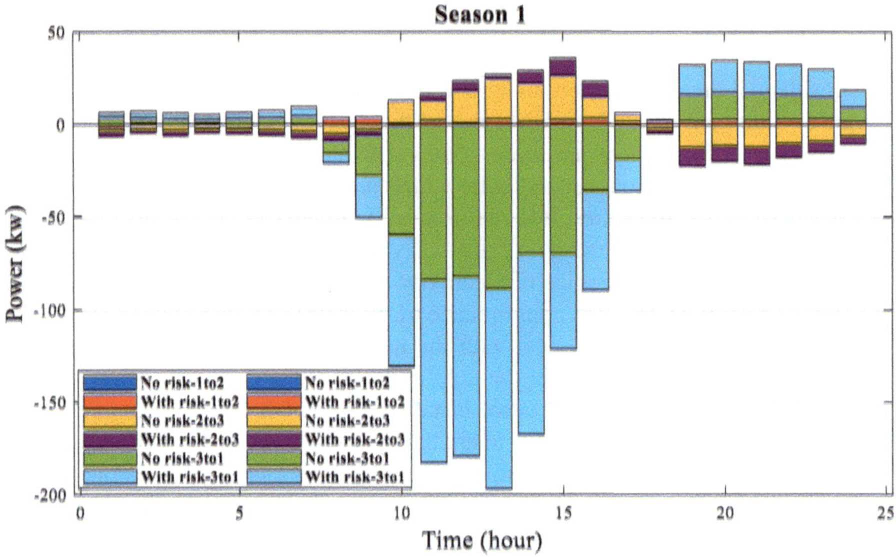


**Fig. 10.2** Traded power between three clients throughout the first season

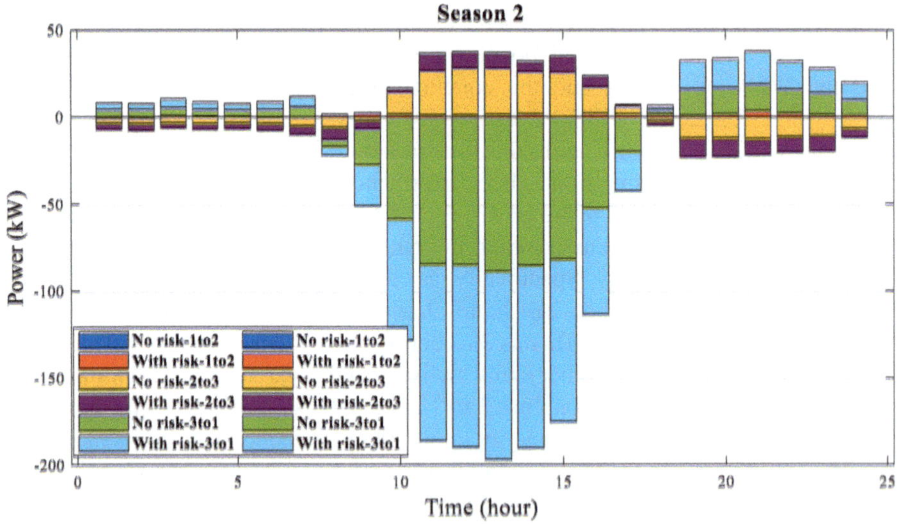


**Fig. 10.3** Traded power between three clients throughout the second season

a noticeable distinction in terms of the amount of power produced, with the With-risk model generating around 140 kilowatts. In addition to this, the amount of traded power in lines 3–1 of the No-risk model is 26% points more than that of the With-risk model.

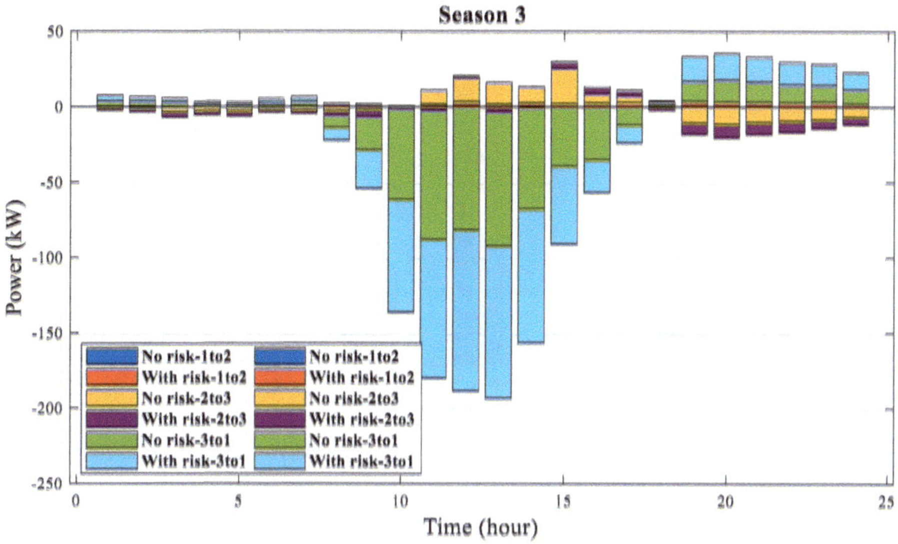


**Fig. 10.4** Traded power between three clients throughout the third season

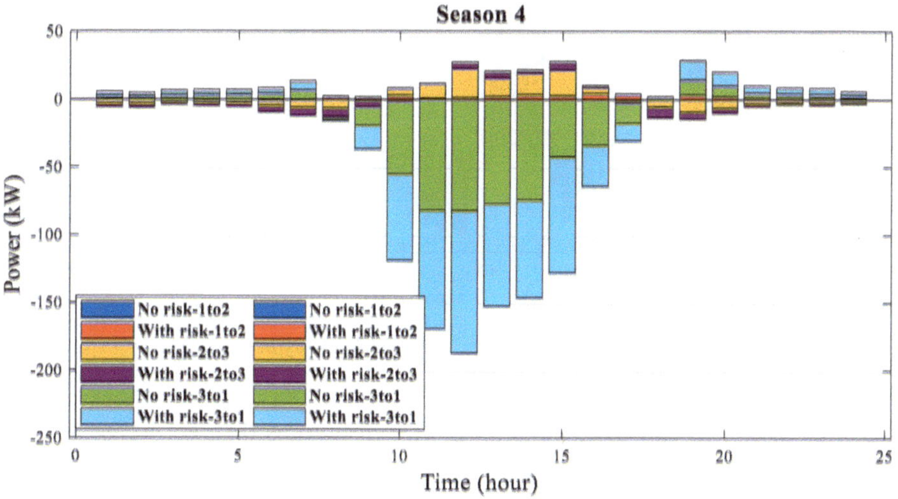


**Fig. 10.5** Traded power between three clients throughout the fourth season

Figure 10.7 depicts the behavior of the hydrogen container while charging/discharging for a year. According to the graph, the bulk of charging has occurred in the middle of the day. This shows that in order to keep hydrogen in the tank, the electrolyzer utilizes the excess energy generated by the solar system. In all models, the charging/discharging methods are equal for the entire year. In the first season division, the peak amount of charged energy in the With/Without-risk models

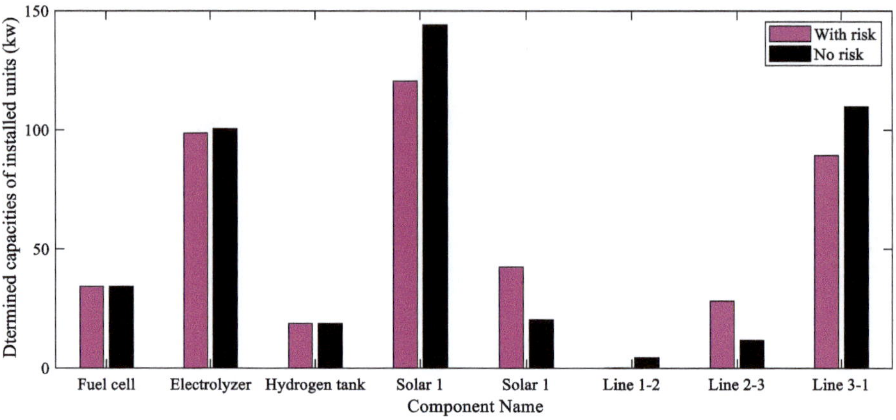


**Fig. 10.6** The performance of the model's components under No-risk and With-risk conditions

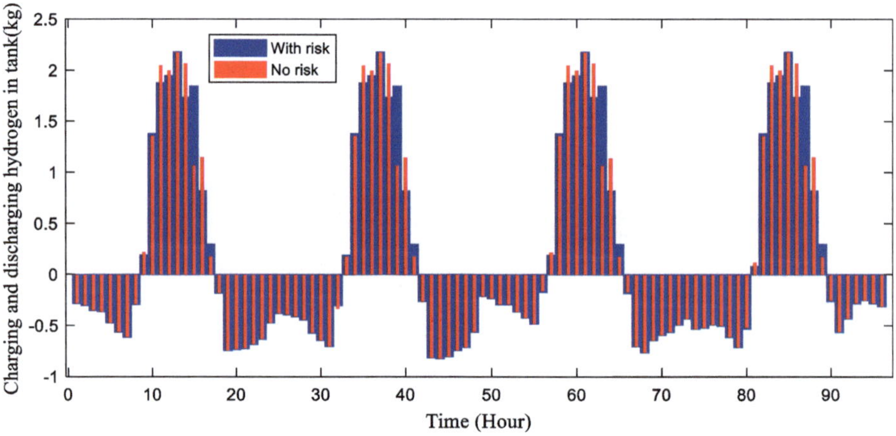


**Fig. 10.7** The level of charged and discharged hydrogen in the hydrogen tank

occurred at age 13 and weighed 2–3 kg. This quantity is 0.75 kg during the first season of the discharge procedure.

In addition, Fig. 10.8 shows the amount of energy stored in the hydrogen tank at different intervals. In the No-risk and With-risk models, this operation has happened throughout the period of 24 hours and year. The system's proficiency is mostly constant across all seasons, however, there are several times when the With/Without-risk models perform somewhat differently. The second season retains somewhat more energy than the previous seasons, with 19 kg. In contrast, the amount of stored energy in the most recent season is 16 kg lower than that of prior seasons.

In addition, Fig. 10.9 displays four seasons in which the operation of the electrolyzer is studied in both With/Without-risk modes. During peak hours (3 p. m.), the With-risk mode electrolyzer uses more energy than the No-risk mode.

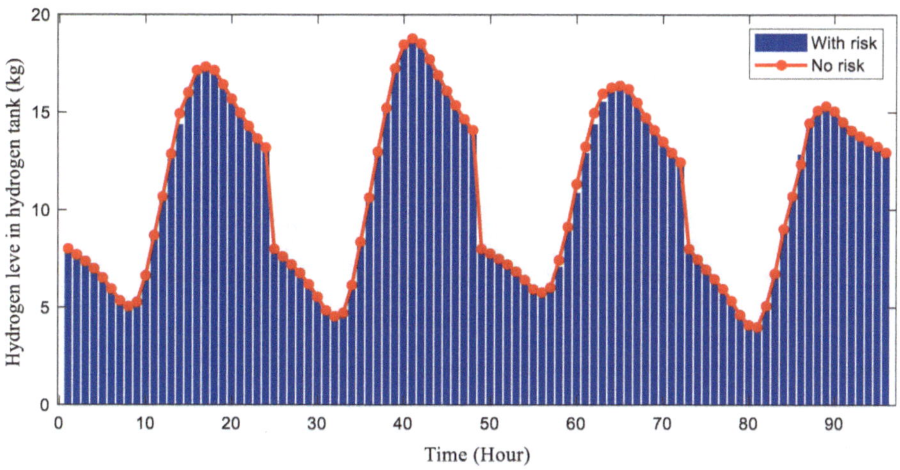


**Fig. 10.8** The amount of hydrogen contained in the tank is both risk-free and risky

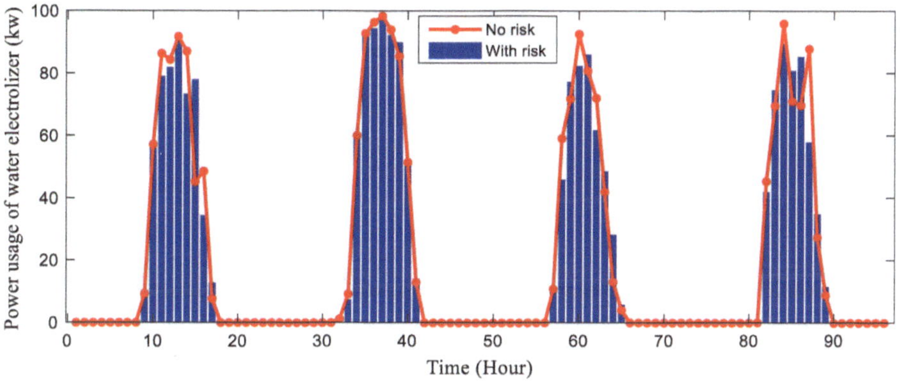


**Fig. 10.9** Four seasons of water electrolyzer-consumed electricity

Moreover, the electrolyzer was mostly active between 9 a.m. and 5 p.m. The second season has the highest power use compared to the previous seasons. Except for seasons 3 and 4, when the No-risk model consumes more power than the With-risk model, the system's performance is almost comparable over all four seasons when using either strategy. During the third season at noon, the No-risk model uses around 95 kw of energy.

In Fig. 10.10, the level of power produced by the fuel cell system is shown to highlight the amount of energy generated at various periods. The quantity of energy utilized by the electrolyzer must match the amount of energy generated by the fuel cell system, according to logic. Clearly, the performance of the electrolyzer and fuel cell systems in this regard is equivalent.

Figure 10.11 is divided into two distinct profiles, illustrating the amount of energy generated by the solar systems of two houses. The first profile in this image

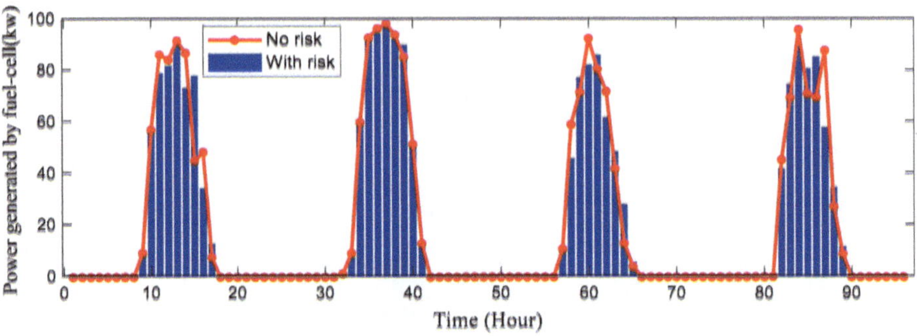


**Fig. 10.10** Four seasons of fuel cell system electricity generation

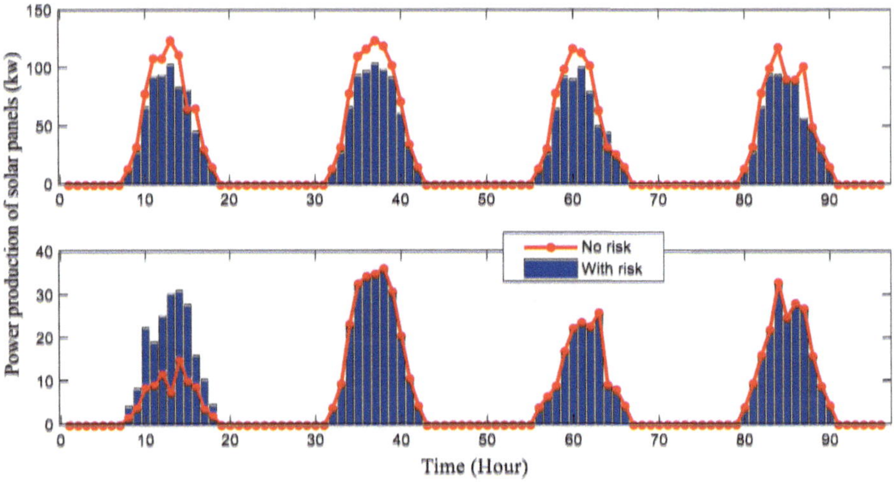


**Fig. 10.11** Solar energy produced by the solar system for two customers throughout four seasons

corresponds to Client 1, while the second profile corresponds to Client 2. Both With/Without-risk models investigate the functioning of the solar system. According to the graph, solar radiation causes the majority of electricity production to occur in the middle of the day. In this regard, owing to the direct radiation of the sun, the quantity of electricity generated (around noon) is larger than at other times. The first profile of the No-risk model produces the maximum power in every season. During the second season, the With-risk model generates around 100 kilowatts of electricity, while the No-risk model restricts output to 125 kilowatts.

In addition to the first season, the With/Without-risk models for Client 2's second profile generate the same amount of electricity. In this manner, the With-risk model generates around 30 kilowatts of initial electricity. The No-risk strategy lowered this quantity to around 12 kilowatts.

## 10.5 Conclusion

The model presented in this study depicts a residential area with three peers, two of which are equipped with PV generation and one with a fuel cell converter system. In contrast, the conditional value-at-risk method was utilized to examine the residential area for both the No/With-risk models. Based on the situation of the customers and the mounting angle of the solar panels with respect to the sun, the results show that the solar systems installed in the two buildings serve different objectives. In addition, based on the actual performance of the first solar system, the No-risk model produces 120 kw more power at noon during the first season than the With-risk model, which only produces 100 kw. In the second solar system, the model with risk produces around 50% more energy than the one without risk. In addition, the amount of hydrogen stored in the tank suggests that the peak energy storage capacity occurred at about 5 p.m. with a value of around 17 kg, which is consistent with both the With/Without-risk models. According to the results of the investigation, the power transmission between buildings 1 and 2 is entirely risk-free. In addition, the With-risk model increases the frequency of transactions between buildings 2 and 3. The No-risk model, in contrast, expects a stronger exchange rate between buildings 3 and 1. During the first season, the greatest amount of electricity exported from building 3 to building 1 occurs between 9 a.m. and 5 p.m., with a maximum of 31,108 kilowatts at 1 p.m.

## References

1. Muhsen, H., Allahham, A., Al-Halhouli, A., Al-Mahmodi, M., Alkhraibat, A., & Hamdan, M. (2022). Business model of peer-to-peer energy trading: A review of literature. *Sustainability, 14*(3), 1616.
2. Soto, E. A., Bosman, L. B., Wollega, E., & Leon-Salas, W. D. (2020). Peer-to-peer energy trading: A review of the literature. *Applied Energy*, 116268.
3. Tushar, W., et al. (2021). Peer-to-peer energy systems for connected communities: A review of recent advances and emerging challenges. *Applied Energy, 282*, 116131.
4. Aminlou, A., Mohammadi-Ivatloo, B., Zare, K., Razzaghi, R., & Anvari-Moghaddam, A. (2022). Peer-to-peer decentralized energy trading in industrial town considering central shared energy storage using alternating direction method of multipliers algorithm. *IET Renewable Power Generation, 16*(12), 2579–2589.
5. Spiliopoulos, N., et al. (2022). Peer-to-peer energy trading for improving economic and resilient operation of microgrids. *Renewable Energy, 199*, 517–535.
6. Reid, F., & Harrigan, M. (2013). An analysis of anonymity in the bitcoin system. In *Security and privacy in social networks* (pp. 197–223). Springer.
7. Zhang, C., Wu, J., Long, C., & Cheng, M. (2017). Review of existing peer-to-peer energy trading projects. *Energy Procedia, 105*, 2563–2568.
8. Sousa, T., Soares, T., Pinson, P., Moret, F., Baroche, T., & Sorin, E. (2019). Peer-to-peer and community-based markets: A comprehensive review. *Renewable and Sustainable Energy Reviews, 104*, 367–378.

9. Mengelkamp, E., Notheisen, B., Beer, C., Dauer, D., & Weinhardt, C. (2018). A blockchain-based smart grid: Towards sustainable local energy markets. *Computer Science-Research and Development, 33*(1–2), 207–214.
10. Long, C., Wu, J., Zhang, C., Thomas, L., Cheng, M., & Jenkins, N. (2018). Peer-to-peer energy trading in a community microgrid. In *IEEE Power and Energy Society general meeting* (Vol. 2018, pp. 1–5). IEEE. https://doi.org/10.1109/PESGM.2017.8274546
11. Zhou, Y., Wu, J., & Long, C. (2018). Evaluation of peer-to-peer energy sharing mechanisms based on a multiagent simulation framework. *Applied Energy, 222*, 993–1022.
12. Mengelkamp, E., Gärttner, J., Rock, K., Kessler, S., Orsini, L., & Weinhardt, C. (2018). Designing microgrid energy markets: A case study: The Brooklyn Microgrid. *Applied Energy, 210*, 870–880.
13. Karami, M., & Madlener, R. (2022). Business models for peer-to-peer energy trading in Germany based on households' beliefs and preferences. *Applied Energy, 306*, 118053.
14. Liu, N., Yu, X., Wang, C., Li, C., Ma, L., & Lei, J. (2017). Energy-sharing model with price-based demand response for microgrids of peer-to-peer prosumers. *IEEE Transactions on Power Systems, 32*(5), 3569–3583.
15. Malik, S., Duffy, M., Thakur, S., Hayes, B., & Breslin, J. (2022). A priority-based approach for peer-to-peer energy trading using cooperative game theory in local energy community. *International Journal of Electrical Power & Energy Systems, 137*, 107865.
16. Zhang, H., Zhang, S., Hu, X., Cheng, H., Gu, Q., & Du, M. (2022, January). Parametric optimization-based peer-to-peer energy trading among commercial buildings considering multiple energy conversion. *Applied Energy, 306*. https://doi.org/10.1016/J.APENERGY.2021.118040
17. Mehdinejad, M., Shayanfar, H., & Mohammadi-Ivatloo, B. (2022). Decentralized blockchain-based peer-to-peer energy-backed token trading for active prosumers. *Energy, 244*, 122713.
18. Belgioioso, G., Ananduta, W., Grammatico, S., & Ocampo-Martinez, C. (2022). Operationally-safe peer-to-peer energy trading in distribution grids: A game-theoretic market-clearing mechanism. *IEEE Transactions on Smart Grid, 13*(4), 2897–2907.
19. Mehrjerdi, H. (2019). Optimal correlation of non-renewable and renewable generating systems for producing hydrogen and methane by power to gas process. *International Journal of Hydrogen Energy, 44*(18), 9210–9219.
20. Mehrjerdi, H. (2020). Peer-to-peer home energy management incorporating hydrogen storage system and solar generating units. *Renewable Energy, 156*, 183–192.
21. Liu, J., Yang, H., & Zhou, Y. (2021). Peer-to-peer energy trading of net-zero energy communities with renewable energy systems integrating hydrogen vehicle storage. *Applied Energy, 298*, 117206.
22. Li, Y., Yu, C., Liu, Y., Ni, Z., Ge, L., & Li, X. (2022). Collaborative operation between power network and hydrogen Fueling stations with peer-to-peer energy trading. *IEEE Transactions on Transportation Electrification, 9*, 1521.
23. Mehrjerdi, H. (2019). Off-grid solar powered charging station for electric and hydrogen vehicles including fuel cell and hydrogen storage. *International Journal of Hydrogen Energy, 44*(23), 11574–11583.
24. Rockafellar, R. T., & Uryasev, S. (2000). Optimization of conditional value-at-risk. *Journal of Risk, 2*, 21–42.

# Index





V. Vahidinasab, B. Mohammadi-Ivatloo (eds.), *Demand-Side Peer-to-Peer Energy Trading*, Green Energy and Technology, https://doi.org/10.1007/978-3-031-35233-1

MIX
Papier aus verantwortungsvollen Quellen
Paper from responsible sources
FSC® C105338

If you have any concerns about our products, you can contact us on
**ProductSafety@springernature.com**

In case Publisher is established outside the EU, the EU authorized representative is:
**Springer Nature Customer Service Center GmbH**
**Europaplatz 3, 69115 Heidelberg, Germany**

Printed by Libri Plureos GmbH
in Hamburg, Germany